AF344283

de Gruyter Studies in Mathematics

Alexander Bendikov

Potential Theory on Infinite-Dimensional Abelian Groups

Translated from the Russian by Carol Regher

Walter de Gruyter
Berlin · New York 1995

Author
Alexander Bendikov
Fakultät für
Mathematik und Informatik
Universität Passau
D-94030 Passau, FRG

Translator
Carol Regher
Physics Department
Kansas State University
Manhattan, Kansas 66506
USA

Series Editors

Heinz Bauer
Mathematisches Institut
der Universität
Bismarckstraße 1½
D-91054 Erlangen, FRG

Jerry L. Kazdan
Department of Mathematics
University of Pennsylvania
209 South 33rd Street
Philadelphia, PA 19104-6395, USA

Eduard Zehnder
ETH-Zentrum/Mathematik
Rämistraße 101
CH-8092 Zürich
Switzerland

1991 Mathematics Subject Classification: 31-02; 60-02

Keywords: Dirichlet problem, harmonic space, Markov process, martingale,
locally compact group, multiplier, hypoellipticity

♾ Printed on acid-free paper which falls within the guidelines of the ANSI to ensure permanence and durability.

Library of Congress Cataloging-in-Publication Data

Bendikov, Alexander, 1950–
[Veroiatnostnaia teoriia potentsiala na beskonechnomernykh
abelevykh gruppakh. English]
Potential theory on infinite-dimensional Abelian groups / Alexander Bendikov; translated from the Russian by Carol Regher
p.   cm. – (De Gruyter studies in mathematics ; 21)
Includes bibliographical references and index.
ISBN 3-11-014283-X (alk. paper)
1. Potential theory (Mathematics) 2. Abelian groups. I. Title.
II. Series.
QA404.7.B4613   1995
516.2'33–dc20   95-14980
                                                                 CIP

Die Deutsche Bibliothek – Cataloging-in-Publication Data

Bendikov, Aleksandr:
Potential theory on infinite dimensional Abelian groups / Alexander
Bendikov. Transl. from the Russ. by Carol Regher. – Berlin ; New
York : de Gruyter, 1995
(De Gruyter studies in mathematics ; 21)
ISBN 3-11-014283-X
NE: GT

Printed in Germany.
Typesetting: Lewis & Leins, Berlin.
Printing: Gerike GmbH, Berlin.
Binding: Fuhrmann KG, Berlin. Cover design: Rudolf Hübler, Berlin.

Contents

Chapter 1
Introduction

Potential theory should long have been considered a chapter of mathematical analysis, using ideas and methods of the theory of functions, functional analysis, and general topology.

An important stage in the development of this theory was opened by the work of Kakutani, Kac, and Doob (1940–1954), which established the close tie of potential theory with the theory of Brownian motion. Since that time, systematic penetration of the methods of probability theory into potential theory has begun.

Probability methods have significantly improved the understanding of some basic ideas of potential theory. Moreover, they have led to a large number of new results. In turn, due to this confluence, probability theory has derived comparable advantages in terms of mathematical apparatus.

The following fact lies at the base of probability interpretations: if $N(x, y) = c_n|x - y|^{2-n}$ is the Newtonian potential kernel on $\mathbb{R}^n$, $n \geq 3$, and $p(t; x, y) = (\sqrt{2\pi t})^{-n} \exp(-|x - y|^2/2t)$ is the transition density of a Wiener process on $\mathbb{R}^n$, then $N(x, y) = \int_0^\infty p(t; x, y)\, dt$. This relation implies the coincidence of the set of non-negative superharmonic functions with respect to the Laplace operator, with the set of excessive functions with respect to a Wiener semigroup. In turn, this property is equivalent to the coincidence of harmonic measures (i.e. measures giving a solution of the Dirichlet problem for the Laplace operator) with the hitting probabilities for a Wiener process. From the noted equivalence, further probabilistic interpretations of results and concepts of classical potential theory have been obtained (such as balayage, fine topology, polar sets, thinness, regular points, etc.). These aspects are completely set forth in the extensive monograph of Doob [53].

Probability interpretations of the classical theory led Doob ([51] and [52]) to study the Dirichlet problem for the heat equation, where he used a combination of analytic and probability methods for this purpose. In addition, Doob constructed axiomatics of the harmonic functions at the basis of which lay the concept of a Markov process. Based on these investigations, as well as on Tautz's earlier investigations [106], Brelot [43] and later also Bauer [1] developed the theory of harmonic spaces. Starting from the basic properties of harmonic functions (the sheaf property, local solvability of the Dirichlet problem, convergence properties) they constructed a theory which includes a wide class of linear second-order differential equations of elliptic and parabolic types. The investigations of Brelot and

Bauer, as well as further axiomatic constructions, are set forth in the extensive monograph of Constantinescu and Cornea [46].

The same period (1954–1963) saw the investigations of Doob [51], [52], Hunt [65], and Dynkin [54] on the theory of Markov processes and their corresponding semigroups of operators, which led to the rise of probabilistic potential theory. This theory in compact form is given in the monograph of Blumenthal and Getoor [40].

The connection between the axiomatic and probabilistic theories was established in 1963 by Meyer [78]; later in 1967 within the framework of a more general axiomatics, this was done by Boboc, Constantinescu, and Cornea [41]. They showed that by analogy with the classical case, to each harmonic space in which the non-negative constants are superharmonic functions, there corresponds a Markov process whose set of excessive functions coincides with the set of non-negative superharmonic functions of the given harmonic space. These investigations, as well as further axiomatic constructions, including nonlocal potential theory, are set forth in the monograph of Bliedtner and Hansen [39].

The present book also treats the enumerated circle of problems. We shall briefly discuss its contents. Chapter 2 is introductory. It lists basic definitions and some theorems from general potential theory which are systematically used in the present book. The survey of Bauer [5] was used in the writing of this chapter, as well as two survey reports done by the author on the XIX School-Colloquium on Probability Theory and Mathematical Statistics (Bakuriani, 1985) and on the IV School-Seminar on the Theory of Random Processes (Preili, 1987) [21].

In the first two sections of Chapter 3 it is shown that for broad conditions imposed on a Markov process, with respect to this process the harmonic functions satisfy the axioms of a harmonic space with a strong convergence property (Brelot or Doob). These convergence properties are deduced from the Harnack inequality (of elliptic (Sec. 3.1) and parabolic (Sec. 3.2) types). Compactness properties for harmonic functions resulting from this inequality are used later in Chapter 5 in the study of weak solutions of infinite-dimensional elliptic equations.

In Secs. 3.3-3.5 projective sequences of harmonic spaces and their projective limits are studied. Such sequences arise, for example, in the study of harmonic sheaves generated by space-homogeneous processes on projective Lie groups (see [63: p. 24]), in particular, on abelian locally compact connected and locally connected groups. Examples show (Sec. 3.3) that in the general case the limit space does not satisfy all the axioms in the definition of a harmonic space. Nevertheless a "good" Markov process (Sec. 3.3) can be associated with this limit space, and thus, a suitable potential theory has been developed (see [18], [104]).

Classical potential theory on $\mathbb{R}^n$ is invariant with respect to translation, and therefore it is natural to consider a harmonic space $(E, \mathcal{H})$ on a topological group E with a translation-invariant sheaf $\mathcal{H}$ of harmonic functions on E. The corresponding theory (harmonic groups) was constructed by Bliedtner [36] in 1968. However, the only source of examples of harmonic groups were translation-invariant elliptic or parabolic second-order differential operators on Lie groups.

In [37] Bliedtner made an attempt to show that the phase space of a harmonic group must be a Lie group, but the proof contained inaccuracies.

In 1972 Forst [55] considered the problem of constructing a harmonic group starting with a weakly continuous semigroup of symmetric probability measures on a locally compact abelian group E. This semigroup of measures induced on E the semigroup of transition probabilities of a certain continuous Hunt process. Based on the results of Port and Stone [88] on space-homogeneous Markov processes, Forst showed that for particular conditions on a semigroup, the harmonic functions determined by the corresponding Hunt process satisfy the axioms of a harmonic group. However, except for the Wiener semigroup on $\mathbb{R}^n$, examples of semigroups satisfying the Forst conditions were not known.

In 1974, in article [7] (see also [6]), the author studied the properties of continuity and compactness of the sheaf $\mathcal{H}$ of harmonic functions of a continuous space-homogeneous process on a group. In particular it was shown that the Forst conditions are not only sufficient but also necessary. In addition, a semigroup of measures was constructed on the group $E = \mathbb{R}^n \times \mathbb{T}^\infty$ satisfying these conditions. Thus, a counterexample to the Bliedtner hypothesis was constructed.

The essence of the matter was the construction of a semigroup of Gaussian measures on a locally compact abelian group E, which are absolutely continuous with respect to Haar measure. Such a construction, as noted in [63: Chapter 5], is possible only under the condition that E is connected and locally connected, i.e. that E is isomorphic to the group $\mathbb{R}^n \times \mathbb{T}^m$ ($m \leq \infty$).

Thus, of all the abelian groups which are not Lie groups, only groups of the type $\mathbb{R}^n \times \mathbb{T}^\infty$ admit a potential theory in the sense of Bliedtner.

In 1976 Berg arrived at a similar construction [30], [32] (see also Bauer [3]). Avoiding probabilistic interpretations, he considered harmonic functions on the group $\mathbb{T}^\infty$ as generalized solutions (uniform limits of "cylindrical" solutions) of a certain second-order elliptic equation of infinitely many variables. Using the probabilistic approach, the author of the present book in 1979 constructed a harmonic group on $\mathbb{R}^n \times \mathbb{T}^\infty$ from a parabolic second-order operator with infinitely many variables and its associated Hunt "heat" process [10].

We note that in the case of $\mathbb{R}^n$ and the heat equation, a similar method was first suggested by Doob [52].

In Chapter 4 the subject of our investigation is potential theory on the infinite-dimensional group $E = \mathbb{R}^n \times \mathbb{T}^\infty$. Thus, in Sec. 4.1, starting from the harmonic group $(E, \mathcal{H})$, we deduce the infinite-dimensional analogue of the well-known Bony theorem [42: Theorem 1.1]. More precisely, we shall show that there exists an infinite-dimensional elliptic differential operator $\mathcal{L} = \sum_{i,j=1}^\infty a_{ij} \partial_i \partial_j + \sum_{i=1}^\infty b_i \partial_i$ such that sheaf $\mathcal{H}$ coincides with the sheaf $\mathcal{H}_{\mathcal{L}}$ of continuous functions that satisfy the equation $\mathcal{L}u = 0$ in the sense of the theory of distributions. In Sec. 4.4 we consider the converse problem: we fix an elliptic differential operator $\mathcal{L}$ and deduce conditions on its coefficients $(a_{ij})_1^\infty$ and $(b_i)_1^\infty$ for which $(E, \mathcal{H}_{\mathcal{L}})$ is a harmonic group. Results obtained in this section can be considered an infinite-dimensional analogue of the second Bony theorem [42: Theorem 5.3, Corollary]. We note that in our investigation of the correspondence "elliptic operator - harmonic sheaf," the connecting link (the same as in classical potential theory) is a space-homogeneous Markov process X on E with almost surely (a.s.) continuous trajectories.

Process X is associated in a natural manner with the semigroup $(\mu_t)_{t>0}$ of Gaussian measures on E, where the necessary properties of harmonic functions (continuity, convergence properties, existence of base of regular sets) are equivalent to specific properties of $(\mu_t)_{t>0}$ (absolute continuity with respect to Haar measure, continuity of density, existence of given asymptotics for $t \downarrow 0$). The construction of semigroups of Gaussian measures on E, the study of their properties, and a proof of the equivalence of these properties to the mentioned property of harmonic functions are given in Secs. 4.2 and 4.3. The results obtained in these sections naturally come into the circle of investigations to which monographs [63] and [29] are dedicated.

In some problems of harmonic analysis [103], [12], [86], [13], as well as in problems of statistical mechanics [64](see also (*) and references mentioned there), the necessity arises for a study of differentiability properties of weak solutions of an elliptic equation on group $\mathbb{T}^\infty$. In Chapter 5 we consider a continuous space-homogeneous process X on $\mathbb{T}^\infty$ with independent coordinates. On smooth functions depending on a finite number of variables, its L_p-generator coincides with differential operator $\mathcal{L} = \sum_{k=1}^\infty a_k \partial_k^2$, $a_k > 0$. In Secs. 5.1 and 5.4, we introduce some *apriori* inequalities which are used for the study of weak solutions of the equation $\mathcal{L}u - \lambda u = f$. In Sec. 5.2 it is shown that every weak solution is infinitely differentiable in L_p ($p > 1$), if the right-hand side of the equation is infinitely differentiable in L_p. In distinction from the finite-dimensional case, this property does not imply the existence of a continuously differentiable (or even simply continuous) regularization. For example, every function that is harmonic on set $V \subset T^\infty$ has generalized derivatives on V in $L_p(W)$, $\overline{W} \subset V$ of any order. However, in the general case (for example, for the infinite-dimensional Laplace operator $\mathcal{L} = \sum_{k=1}^\infty \partial_k^2$), there exist discontinuous harmonic functions. In Sec. 5.3 it is shown that the existence of a continuously differentiable regularization is connected with the Harnack inequality, which holds if the coefficients a_k of differential operator $\mathcal{L}$ increase sufficiently quickly for $k \uparrow \infty$.

Chapter 6 consists of two parts. The first part (Sec. 6.1) introduces the mixed norm on a set of martingales with respect to a special family of σ-algebras. Such spaces of martingales (denoted by $\mathcal{M}_{\vec{p}}$) naturally arise during transference of the definition of mixed norm from finite-dimensional structures [34] to infinite-

---

(*) S. Abeverio, Y.G. Kondratiev and M. Röckner, Infinite Dimensional Diffusion, Markov Fields, Quantum Fields and Stochastic Quantization, in: Stochastic Analysis and Applications in Physics, NATO ASI Series, Ser. C: Math. & Phys. Sci., Vol. 449, Kluwer Acad. Publ. (1994)

dimensional ones [17]. Many properties of spaces of martingales $\mathcal{M}_{\vec{p}}$ are analogous to properties of ordinary spaces L_p of integrable functions, where these spaces are identified if $\vec{p} = (p, p, \ldots)$ and $p > 1$. At the same time, as shown in the second part (Secs. 6.2 and 6.3), the scale of spaces $\{\mathcal{M}_{\vec{p}}\}$ proves to be more suited to some problems of potential theory on group $\mathbb{T}^\infty$ than the scale of ordinary spaces $\{L_p\}$. Thus, in Sec. 6.2, considering harmonic functions on semispace $\mathbb{T}^\infty_+ =]0, +\infty[\times \mathbb{T}^\infty \subset \mathbb{R}^1 \times \mathbb{T}^\infty$, we use the $\mathcal{M}_{\vec{p}}$-norm to study the boundary behavior of these functions, as well as for the problem of representing them in the form of a Poisson integral. Classical finite-dimensional results on this problem are given in monograph [102: Chapter 2].

In Sec. 6.3 we use the $\mathcal{M}_{\vec{p}}$-norm to study the Gaussian semigroups of measures and Bessel potentials generated by them. This approach is dictated by the following circumstance. If $(\mu_t)_{t>0}$ is a Gaussian semigroup and $(\mathcal{J}_\alpha)_{\alpha>0}$ is its associated family of Bessel potentials, then in distinction from the finite-dimensional case (see [101: Chapter 5, Sec. 3]), it can happen that for all $t, \alpha > 0$, measures μ_t and $\mathcal{J}_\alpha$ are singular with respect to Haar measure (in this connection $\operatorname{supp}\mu_t = \operatorname{supp}\mathcal{J}_\alpha = \mathbb{T}^\infty$). This situation takes place, for example, in the case of the Gaussian semigroup associated with the Laplace operator $\mathcal{L} = \sum_{k=1}^\infty \partial_k^2$ on $\mathbb{T}^\infty$. Moreover, even in the case of absolute continuity, $\mathcal{J}_\alpha$ is not in L_p for any $p > 1$. At the same time, for all $t > 0$, measures μ_t are in $\mathcal{M}_{\vec{p}}$, if components p_k of vector $\vec{p}$ go to the identity sufficiently quickly for $k \uparrow \infty$. In addition, for $t \downarrow 0$, we have $\|\mu_t\|_{\mathcal{M}_{\vec{p}}} \sim t^{-\beta}$, where $\beta = \sum_{k=1}^\infty (1 - \frac{1}{p_k})$; therefore also $\mathcal{J}_\alpha \in \mathcal{M}_{\vec{p}}$ for all $\alpha > \beta$. Thus we arrive at the infinite-dimensional analogue of Sobolev's inequality [101: p. 141]:

$$\|\mathcal{J}_\alpha * f\|_{\mathcal{M}_{\vec{q}}} \leq c\|f\|_{\mathcal{M}_{\vec{p}}},$$

$$1/q_k = 1/p_k - 1/\Delta_k \ (k \geq 1), \ \sum_{k=1}^\infty 1/\Delta_k < \alpha.$$

We note that Sobolev's inequality in mixed norms on group $\mathbb{R}^n$ has been known a rather long time [34]. However, the method of proof of this inequality gives a constant $c = c(n)$ which increases exponentially for $n \uparrow \infty$. This circumstance does not permit one to obtain the infinite-dimensional result from the finite-dimensional using the method of periodization [102: Chapter 7] and the procedure of passing to the limit for $n \uparrow \infty$.

From the Sobolev inequality, as in the finite-dimensional case [101: Chapter 5, Sec. 2.2], one can deduce that the space of Bessel potentials $L_{\vec{p}}^\alpha = \{\mathcal{J}_\alpha * \varphi : \varphi \in L_{\vec{p}}\}$ is included in the space of continuous functions $\mathbb{C}(\mathbb{T}^\infty)$.

The author would like to express sincere thanks to Professors N. S. Landkof, A. N. Shiryaev, and H. Bauer for the constant attention and valuable advice which they gave the author in the process of work on this book in Russia and in Germany. We thank Mrs. C. Regehr for her translation of the manuscript. Thanks are due to Dr. M. Karbe of Walter de Gruyter & Co. for his constant and truly generous cooperation. We also owe thanks to the Alexander von Humboldt Foundation for enabling us to spend the years 1992–1994 in Erlangen, where the project took its final form under the stimulating influence of our German colleagues.

Erlangen, 1994   *Alexander Bendikov*

Chapter 2
Elements of potential theory

In this section we introduce some basic facts from general potential theory which are systematically used in the present work. On the whole we shall follow the monographs of Constantinescu and Cornea [46] and Bliedtner and Hansen [39]. In problems touching on classical theory, we keep to the monograph of Doob [53].

2.1 Notation

Let E be a locally compact Hausdorff space with countable base. By $\mathcal{U}$ (respectively $\mathcal{U}_k$) we shall denote the class of open (respectively open with compact closure) subsets of E. If $A \subset E$, then $\overline{A}$ and ∂A will denote the closure and the boundary of A.

$\mathcal{B}$ is the σ-algebra of Borel subsets of E.

$\mathbb{B}$ (respectively $\mathbb{B}_k$) is the linear space of bounded Borel functions (respectively bounded Borel functions with compact support) on E with norm $\|f\| = \sup_{x\in E} |f(x)|$. $\mathbb{C}$ is the linear space of continuous functions on E; $\mathbb{C}_b = \mathbb{C} \cap \mathbb{B}$, $\mathbb{C}_k = \mathbb{C} \cap \mathbb{B}_k$ and $\mathbb{C}_0$ is the closure of $\mathbb{C}_k$ in $\mathbb{B}$.

If μ is a measure on $(E, \mathcal{B})$ and $f, g \in \mathcal{B}$, then for integral $\int f \, d\mu$ we shall also use the notation $\mu(f)$ or $\langle \mu, f \rangle$; in addition, $\langle f, g \rangle_\mu = \int f g \, d\mu$. Sometimes in the notation $\langle f, g \rangle_\mu$ we shall omit the index "μ."

If $N(x, \cdot)$ $(x \in E)$ is a measurable family of measures on $(E, \mathcal{B})$, we shall write

$$\mu N(\cdot) := \int N(x, \cdot) \, d\mu(x).$$

2.2 Harmonic and hyperharmonic sheaves

A sheaf of functions on E is a map $\mathcal{H}$ defined on $\mathcal{U}$ such that:

1) for any $V \in \mathcal{U}$, $\mathcal{H}(V)$ is a set of functions on V;

2) for any $V_1, V_2 \in \mathcal{U}$ such that $V_1 \subset V_2$, the restriction of any function in $\mathcal{H}(V_2)$ to V_1 belongs to $\mathcal{H}(V_1)$;

3) for any family $(V_i) \subset \mathcal{U}$, a function on $V = \cup_i V_i$ belongs to $\mathcal{H}(V)$ if for each i its restriction to V_i belongs to $\mathcal{H}(V_i)$.

The sheaf $\mathcal{H}$ is called a harmonic sheaf if for any $V \in \mathcal{U}$, $\mathcal{H}(V)$ is a linear space of real continuous functions on V. Functions in $\mathcal{H}(V)$ are called harmonic functions on V.

A set $V \in \mathcal{U}_k$ is called regular (with respect to $\mathcal{H}$) if any continuous function f on ∂V admits a unique continuous continuation H_f to $\overline{V}$ that is harmonic on V and non-negative if f is non-negative.

For a regular set V and any $x \in V$, the map $f \to H_f(x)$ is a positive linear functional defining a measure μ_x^V concentrated on ∂V. The measure μ_x^V is called the harmonic measure of set V at point x.

It is said that a harmonic sheaf $\mathcal{H}$ is nondegenerate at a point x if in some neighborhood of x there exists a harmonic function u such that $u(x) \neq 0$.

It is said that harmonic sheaf $\mathcal{H}$ admits:

1) *the Bauer convergence property* [1] if for any set $V \in \mathcal{U}$ the limit of any increasing sequence of harmonic functions on V is a harmonic function if it is locally bounded;

2) *the Doob convergence property* [52] if for any set $V \in \mathcal{U}$ the limit of any increasing sequence of harmonic functions on V is a harmonic function if it is finite on a dense set;

3) *the Brelot convergence property* [43] if for any connected set $V \in \mathcal{U}$ the limit of any increasing sequence of harmonic functions on V is a harmonic function if it is finite at least at one point.

A sheaf of functions $\mathcal{S}$ on E is called a hyperharmonic sheaf if $\mathcal{S}(V)$ is a convex cone of lower-semicontinuous, $]-\infty, +\infty]$-valued functions on V for any $V \in \mathcal{U}$. The map $V \to \mathcal{S}(V) \cap (-\mathcal{S}(V))$ is obviously a harmonic sheaf, which is denoted by $\mathcal{H}_{\mathcal{S}}$.

The set $V \in \mathcal{U}$ is called an MP-set (min principle) if any function $f \in \mathcal{S}(V)$, non-negative outside the intersection of V with some compact set $K \subset E$ and for which $\lim\inf f \geq 0$ at every boundary point, is non-negative on V.

2.3 The generalized Dirichlet problem

Let V be an MP-set and f a numerical function on ∂V. We denote by $\mathcal{S}_f$ the set of functions $u \in \mathcal{S}(V)$ which are bounded below on V, non-negative outside the intersection of V with some compact set and such that for any $y \in \partial V$

$$\liminf_{x \to y} u(x) \geq f(y).$$

We set $\overline{H}_f = \inf\{u : u \in \mathscr{S}_f\}$ and $\underline{H}_f = -\overline{H}_{(-f)}$. We have $\underline{H}_f \leq \overline{H}_f$. In the case when functions $\overline{H}_f$ and $\underline{H}_f$ are equal and belong to $\mathscr{H}_{\mathscr{S}}(V)$, the function f is called resolutive, and the common value of $\overline{H}_f$ and $\underline{H}_f$ is called the generalized solution of the Dirichlet problem with boundary function f and is denoted by H_f.

The set $V \in \mathscr{U}$ is called resolutive if any function $f \in \mathbb{C}_k$ is resolutive. For a resolutive set V and any $x \in V$, the map $f \to H_f(x)$ is a positive linear functional defining a measure μ_x^V on ∂V. The measure μ_x^V is called the harmonic measure of V at point x.

A point $y \in \partial V$ is called regular if for any function $f \in \mathbb{C}_k$:

$$\lim_{x \to y} H_f(x) = f(y).$$

A resolutive set is called regular if all its boundary points are regular.

2.4 Harmonic spaces

A space E endowed with a hyperharmonic sheaf $\mathscr{S}$ is called a harmonic space if the following axioms are satisfied:
1) (axiom of nondegeneracy) the sheaf $\mathscr{H}_{\mathscr{S}}$ is nondegenerate at all points $x \in E$;
2) (axiom of convergence) $\mathscr{H}_{\mathscr{S}}$ admits the Bauer convergence property;
3) (axiom of resolutivity) the resolutive sets form the base of a topology of E;
4) (axiom of completeness) a lower-semicontinuous function u $(-\infty < u \leq +\infty)$ on a set $V \in \mathscr{U}$ belongs to $\mathscr{S}(V)$ if for any resolutive set $W : \overline{W} \subset V$ and for any $x \in W$:

$$\mu_x^W(u) \leq u(x).$$

A harmonic space E is locally connected and has no isolated points ([46], p. 31).

A function $u \in \mathscr{S}(V)$ is called superharmonic on set V if for any resolutive set $W \in \mathscr{U}_k$, $\overline{W} \subset V$, the function $x \to \mu_x^W(u) \in \mathscr{H}_{\mathscr{S}}(W)$.

A function $p \geq 0$ that is superharmonic on the whole space E is called a potential if any non-negative harmonic minorant of it is identically zero. The smallest closed set outside of which the potential p is harmonic is called the harmonic support of p and is denoted $S(p)$.

A harmonic space is called a $\mathscr{P}$-space if for any $x \in E$ there exists a potential p such that $p(x) > 0$. In a $\mathscr{P}$-harmonic space every open subset is an MP-set ([46], Corollary 2.3.3.).

2.5 Brelot and Bauer spaces

In earlier axiomatics Brelot [43] and Bauer [1] started with a space E endowed with a harmonic sheaf $\mathcal{H}$. Then the hyperharmonic sheaf $\mathcal{H}^*$ was constructed from sheaf $\mathcal{H}$. We shall dwell briefly on these constructions (see [46], Chapter 3).

A locally compact space E endowed with a harmonic sheaf $\mathcal{H}$ is called a Brelot space if it satisfies the following axioms:

1) E is locally connected and has no isolated points;
2) the regular sets with respect to $\mathcal{H}$ form the base of a topology;
3) $\mathcal{H}$ admits the Brelot convergence property.

Solutions of a wide class of second-order differential equations of elliptic type in a given domain of Euclidean space satisfy the axioms of Brelot space (see [46], Exercises 3.2.7-3.2.10, [73], [39], [53], [4]). In Chapters 4, 5, and 7 we shall consider infinite-dimensional differential equations.

In order to give applications to the theory of equations of parabolic type, Bauer weakened the previous system of axioms in the following manner. A locally compact space E endowed with a harmonic sheaf $\mathcal{H}$ is called a Bauer space if the following axioms are satisfied:

1) $\mathcal{H}$ is non-degenerate at every point $x \in E$;
2) there exists a base $\mathcal{F}$ of regular sets with respect to $\mathcal{H}$ such that for any $V_1, V_2 \in \mathcal{F}$ the set $V_1 \cap V_2$ is regular with respect to $\mathcal{H}$;
3) $\mathcal{H}$ admits the Bauer convergence property.

For applications to equations of parabolic type, see [46] (Exercises 3.3.5–3.3.7), [73], [39], and [53]. Applications to infinite-dimensional equations will be given in Chapters 4, 5, and 7.

A hyperharmonic sheaf $\mathcal{H}^*$ in Brelot and Bauer spaces is defined in the following manner. The set $\mathcal{H}^*(V)$, $V \in \mathcal{U}$, consists of the lower-semicontinuous functions h $(-\infty < h \leq +\infty)$ on V which satisfy the condition: for any point $x_0 \in V$ there exists a neighborhood V_0 of it such that for any $W \in \mathcal{F} : \overline{W} \subset V_0$ and $x \in W$ we have $\mu_x^W(h) \leq h(x)$, where $\mu_{\cdot}^W$ is the harmonic measure determined by $\mathcal{H}$.

Every Brelot space is a Bauer space ([46], Theorem 3.1.2). For any Bauer space $(E, \mathcal{H})$ the space $(E, \mathcal{H}^*)$ is a harmonic space, and $\mathcal{H}_{\mathcal{H}^*} = \mathcal{H}$ ([46], Corollary 3.1.1). Finally we note that if in harmonic space $(E, \mathcal{S})$ the family of regular sets forms the base of a topology, then $(E, \mathcal{H}_{\mathcal{S}})$ is a Bauer space, and $(\mathcal{H}_{\mathcal{S}})^* = \mathcal{S}$ ([46], Corollary 3.1.2).

The most general criterion of regularity of a boundary point x_0 of set $V \in \mathcal{U}$ is the criterion based on the concept of a barrier ([54], Theorem 1.3.6; [46], Corollary 2.4.3). A function f defined on some neighborhood W of point x_0 is called a barrier for (x_0, V) if $f > 0$, $f \in \mathcal{S}(W \cap V)$,

$$\lim_{x \to x_0, \, x \in V \cap W} f(x) = 0.$$

Regularity criterion: a boundary point x_0 of set $V \in \mathcal{U}$ is regular if and only if a barrier exists for (x_0, V) .

2.6 Smooth Bauer spaces

For applications it is important to determine all sheaves satisfying the axioms. For the case of a domain in $\mathbb{R}^n$ and functions of class $\mathbb{C}^2$, as well as for the case of the whole space $\mathbb{R}^n$ and sheaves that are translation-invariant and contain constants, this was done by Bony [42] with the help of a pre-elliptic (possibly degenerate) operator. Here, following [39: Chap. 8], we introduce Bony's results.

Let $E \subset \mathbb{R}^n$ be a domain and $\mathcal{H}$ a harmonic sheaf on E such that $(E, \mathcal{H})$ is a Bauer space. It is said that $(E, \mathcal{H})$ is a smooth Bauer space if $\mathcal{H}(V) \subset \mathbb{C}^\infty(V)$ for any $V \in \mathcal{U}$. For simplicity, it is further assumed that $1 \in \mathcal{H}(E)$.

Theorem 2.6.1. *Let $(E, \mathcal{H})$ be a smooth Bauer space. Then there exists a differential operator $\mathcal{L} = \sum_{i,j=1}^{n} a_{ij}\partial_i\partial_j + \sum_{i=1}^{n} b_i\partial_i$ such that:*
1) $(a_{ij})_1^n \geq 0 \ (\not\equiv 0)$;
2) $\mathcal{H} \subset \mathcal{H}_\mathcal{L} := \{h \in \mathbb{C}^2 : \mathcal{L}h = 0\}$;
3) *There exists an open dense subset $E_0 \subset E$ such that:*
   a) $a_{ij}\,|_{E_0} \in \mathbb{C}^\infty(E_0) \ (\forall i, j \leq n)$;
   b) *sheaves $\mathcal{H}$ and $\mathcal{H}_\mathcal{L}$ restricted to E_0 coincide.*

Sheaves $\mathcal{H}$ and $\mathcal{H}_\mathcal{L}$ in the general case do not coincide on E; the coincidence will take place if coefficients (a_{ij}) are continuous everywhere (not only on E_0). We give an example.

Example. Let $E = \mathbb{R}^1$ and for every open interval V let $\mathcal{H}(V)$ be the set of functions on V of the type $x \to \alpha x^3 + \beta$ $(\alpha, \beta \in \mathbb{R}^1)$. We set $E_0 = \mathbb{R}^1 \backslash \{0\}$ and

$$\mathcal{L} := \partial_x^2 - a(x)\partial_x,$$

where $a(x) = \frac{2}{x}$, if $x \neq 0$ and $a(0)$ is any number, for example equal to 1. It is obvious that the operator $\mathcal{L}$ and the set E_0 are those needed in the sense of Theorem 2.6.1. Now we note that the function

$$u(x) = \left\{ \begin{array}{ll} x^3, & x \leq 0 \\ 2x^3, & x > 0 \end{array} \right.$$

belongs to $\mathcal{H}_\mathcal{L}(E)$, but does not belong to $\mathcal{H}(E)$.

We note that the differential operator $\mathcal{L}$ is determined by sheaf $\mathcal{H}$ uniquely up to multiplication by a function of class $\mathbb{C}^\infty$ ([39], Chap. 8, Sec. 2.7).

The condition $\mathcal{H} \subset \mathbb{C}^\infty$ in Theorem 2.6.1 can be replaced by a weaker condition [42]: any harmonic function can be locally uniformly approximated by harmonic functions of class $\mathbb{C}^\infty$. If $E = \mathbb{R}^n$ and sheaf $\mathcal{H}$ is translation-invariant, then this condition is satisfied. As a supplement to parts 1–3 of Theorem 2.6.1, we shall describe this special case, which is important for us ([42], Theorem 3.4):

4) There exists a differential operator $\mathcal{L} = \sum_{i,j=1}^\infty a_{ij}\partial_i\partial_j + \sum_{i=1}^\infty b_i\partial_i$ with constant coefficients such that $\mathcal{H} = \mathcal{H}_{\mathcal{L}}$. Moreover $(E, \mathcal{H})$ is a Brelot space if and only if $\mathrm{rang}(a_{ij})_1^n = n$.

Theorem 2.6.2. *Let $\{A_1, ..., A_r, A_0\}$ be $\mathbb{C}^\infty$-vector fields on E. We set*

$$\mathcal{L} := \sum_{k=1}^r A_k^2 + A_0, \quad \mathcal{H} := \mathcal{H}_{\mathcal{L}}.$$

Then $(E, \mathcal{H})$ is a smooth Bauer space if:

$$\mathrm{rang}_x Lie(A_1, ..., A_r, A_0) = n \quad (\forall x \in E);$$

moreover in $(E, \mathcal{H})$ the Doob convergence property is satisfied. If the following condition is satisfied:

$$\mathrm{rang}_x Lie(A_1, ..., A_r) = n \quad (\forall x \in E),$$

then in $(E, \mathcal{H})$ the Brelot convergence property is satisfied.

In Sections 5.1, 5.4, and 6.3 we shall obtain generalizations of these results in the case of differential operators of infinitely many variables.

2.7 Markov processes

We follow the usual terminology for the theory of Markov processes [54], [40]. $X = (x_t, \xi, \mathcal{M}_t, P^x)$ denotes a standard Markov process on measurable space $(E, \mathcal{B})$. We set $P_t f(x) = M_x f(x_t)$, as well as $R^\lambda f(x) = \int_0^\infty e^{-\lambda t} P_t f(x)\, dt$, if the latter integral converges ($\lambda \geq 0$); here $f \in \mathbb{B}$ and M_x is the mathematical expectation with respect to the measure P^x. By the Markov property, $(P_t)_{t>0}$ is a semigroup of the operators in $\mathbb{B}$, and $(R^\lambda)_{\lambda \geq 0}$ is its resolvent. A subprocess of the original process corresponding to the semigroup $\{P_t^\lambda; t > 0\}$, where $P_t^\lambda := e^{-\lambda t} P_t$, is denoted by X^λ.

A universally measurable function $f \geq 0$ is called a λ-supermedian if for any $t > 0$ we have $P_t^\lambda f \leq f$; if furthermore $\lim_{t\to 0} P_t^\lambda f = f$, then the function f is called λ-excessive. For any $\lambda \geq 0$ and universally measurable function $g \geq 0$, the function $R^\lambda g$ is λ-excessive; it is called the λ-potential of the function g.

For the set $A \subset E$ we set

$$T_A = \inf\{t > 0 : x_t \in A\},$$
$$D_A = \inf\{t \geq 0 : x_t \in A\}.$$

According to Hunt's fundamental theorem for a set $A \in \mathcal{B}$ (more generally, almost Borel), T_A and D_A are stopping times.

The point $x \in E$ is called regular (respectively irregular) for an almost Borel set $A \subset E$ if: $P^x(T_A = 0) = 1$ (respectively $P^x(T_A = 0) = 0$). By the "0-1" law, the probability $P^x(T_A = 0)$ can take only the two values 0 or 1; therefore the point x is either regular or irregular for A. By A^r (respectively A^i) we denote the set of all regular (irregular) points of set A; A^r and A^i again are almost Borel sets.

The set $A \subset E$ is called:

a) *thin at a point* $x \in E$ if it is contained in an almost Borel set $D : x \in D^i$;
b) *fine open* if the set $E \backslash A$ is thin at all points of set A; such sets generate a topology on E which is called the *fine topology* (or natural topology [54]) on E generated by process X. In this topology all λ-excessive functions are continuous. It is obvious that the fine topology is finer than the original topology on E.
c) *polar* if it is contained in an almost Borel set $D : P^x(T_D < +\infty) = 0$ $(\forall x \in E)$;
d) *thin* if it is contained in an almost Borel set $D : D^r = \emptyset$. Any countable union of thin sets is called a *semipolar set*; the trajectory $t \to x_t$ of process X. hits this set A no more than countably many times.

The complement of a semipolar set is fine dense in E, and therefore dense in E with respect to the usual topology. A typical situation in which a semipolar set arises is the following: for every almost Borel set A, the set $A \backslash A^r$ is semipolar. We note further that every polar set is semipolar, while the converse statement generally speaking is not true. Thus, for example, if X is the process of uniform motion to the left, then the set $A = \{0\}$ is semipolar, but not polar.

General conditions for which the mentioned classes of sets coincide are listed by Silverstein [98]. In particular, these conditions imply that if X is a symmetric process (i.e. equivalent to its dual process; see Sec. 2.10), then coincidence of these classes of sets takes place. In particular, this coincidence takes place if X is a Wiener process.

The following kernels play an important role in probabilistic potential theory:

$$P_A^\lambda(x, \Gamma) = M_x(e^{-\lambda T_A}; x_{T_A} \in \Gamma),$$
$$\widetilde{P}_A^\lambda(x, \Gamma) = M_x(e^{-\lambda D_A}; x_{D_A} \in \Gamma) \quad (x \in E, \Gamma \in \mathcal{B})$$

The characterization of these kernels in terms of harmonic space is the key to the probabilistic interpretation of axiomatic potential theory.

To conclude this section, we give the definition of a λ-superharmonic (λ-harmonic) function with respect to Markov process X. In this connection we shall assume that the trajectories $x.$ are almost surely (a.s.) continuous.

A function f defined on $V \in \mathcal{U}$ and taking values in the interval $]-\infty, +\infty]$ is called λ-superharmonic ($\lambda \geq 0$) on V if the following properties are satisfied:

a) f is almost Borel and fine continuous;
b) f is locally bounded from below;
c) $\forall W \in \mathcal{U} : \overline{W} \subset V$, $\forall x \in W$, we have $P_{E \backslash W}^\lambda(x, f) \leq f(x)$.

The function f is called λ-harmonic on $V \in \mathcal{U}$ if functions f and $-f$ are λ-superharmonic on V.

Every λ-excessive function being restricted to any set $V \in \mathcal{U}$, is λ-superharmonic on V, furthermore if it has the form $R^\lambda g$ ($g \in \mathbb{B}^+$, $\lambda \geq 0$), then it is λ-harmonic on the set $E \backslash \operatorname{supp} g$. Another important example of a λ-harmonic function is the function $P^\lambda_{E\backslash V}(x, f)$ ($V \in \mathcal{U}$, $x \in V$, $f \in \mathbb{B}$).

By the assumption on continuity of trajectories of process X, the concept of λ-superharmonic (λ-harmonic) functions is localized. In fact, let X_V be the part of process X on set V, i.e. a Markov process determined by the semigroup $\{P^V_t ; t > 0\}$:

$$P^V_t f(x) = M_x(f(x_t); t < T_{E\backslash V}).$$

Then for any open set $W \subset V$, the classes of λ-superharmonic (λ-harmonic) functions on W for processes X and X_V coincide. In particular, every function of the type $R^\lambda_V f(x) = M_x \int_0^{T_{E\backslash V}} e^{-\lambda t} f(x_t)\, dt$ ($x \in V$, $f \in \mathbb{B}^+$) is λ-superharmonic on V.

If $\lambda = 0$, then in notation of the type P^λ_t, R^λ, and in the terms "λ-superharmonic (λ-harmonic) function" etc., we shall omit the index λ.

2.8 Markov processes on harmonic spaces

In this section we present two theorems connecting potential theory on a harmonic space with probabilistic potential theory. The first of these belongs to the cycle of theorems "harmonic space $\to$ Markov process" [46; Chapter 10] (see also [78], [2], [5], [39: Chaps. 4, 8.1]), and the second belongs to the cycle of theorems "Markov process $\to$ harmonic space" [38] (see also [51], [108], [18]).

Theorem 2.8.1. *Let $(E, \mathcal{S})$ be a $\mathcal{P}$-harmonic space in which the function $f \equiv 1$ is superharmonic. There exists a standard process $X = (x_t, \xi, \mathcal{M}_t, P^x)$ with almost surely continuous trajectories $x.$ such that the cone of its excessive functions coincides with the cone $\mathcal{S}^+(E)$ of non-negative hyperharmonic functions on E.*

For all spaces $(\mathbb{R}^n, \mathcal{S}_\Delta)$ ($n \geq 3$, $\Delta = \sum_{k=1}^n \partial_k^2$) of classical potential theory, X is a Wiener process. For spaces $(\dot{\mathbb{R}}^n, \mathcal{S}_{\dot\Delta})$ ($\dot{\mathbb{R}}^n = \mathbb{R}^n \times \mathbb{R}^1$, $\dot{\Delta} = \Delta_x - \partial_t$) of parabolic potential theory, the corresponding process is a space-time Wiener process (see [66], [52], [53]).

Now let $X = (x_t, \xi, \mathcal{M}_t, P^x)$ be a standard process with almost surely continuous trajectories $x..$ A function f defined on $V \in \mathcal{U}$ is called a locally superharmonic function if each point $x \in V$ has a neighborhood V_x on which f is superharmonic. We denote the set of locally superharmonic functions on V by $\mathcal{S}(V)$. It is obvious that the map $\mathcal{S} : V \to \mathcal{S}(V)$ is a sheaf.

Theorem 2.8.2. *Let the process X satisfy the conditions:*
1) *for any point $x \in E$ we have $P^x(T_{E\setminus\{x\}} < \xi) > 0$;*
2) *for any compact set $K \subset E$ and any point $x \in E$ we have $R1_K(x) < \infty$;*
3) *every excessive function is the limit of an increasing sequence of continuous excessive functions.*
Then $\mathcal{S}$ is a hyperharmonic sheaf and $(E, \mathcal{S})$ is a $\mathcal{P}$-harmonic space.

2.9 Probability interpretations

Now we fix a $\mathcal{P}$-harmonic space $(E, \mathcal{S})$ and standard process $X = (x_t, \xi, \mathcal{M}_t, P^x)$ with continuous trajectories $x.$ associated with $(E, \mathcal{S})$ in the sense of Theorem 2.8.1 of the preceding section.

For a function $u \in \mathcal{S}^+(E)$ and a set $A \subset E$, we set

$$R_u^A(x) := \inf\{v(x) : v \in \mathcal{S}^+(E), \ v \geq u \cdot 1_A\}.$$

The function R_u^A is called the *reduction* of function u on set A. We have

$$R_u^A \leq u, \ R_u^A(x) = u(x) \ (\forall x \in A).$$

In the general case the function R_u^A is not lower-semicontinuous and therefore is not hyperharmonic. We denote by $\underset{+}{R}{}_u^A$ the lower-semicontinuous regularization of function R_u^A. This function is hyperharmonic (and harmonic on the set $\overset{o}{E\setminus A}$, i.e. the interior of set $E\setminus A$); it is called the *balayage of function u* on set A.

Let μ be a measure on $(E, \mathcal{B})$ with compact support. According to [46] there exists a unique measure μ^A on $(E, \mathcal{B})$ such that

$$\int \underset{+}{R}{}_u^A \, d\mu = \int u \, d\mu^A \ (\forall u \in \mathcal{S}^+(E));$$

the measure μ^A is called the *balayage* of μ on set A.

The operation of balayage of measures is closely connected with the generalized Dirichlet problem. Thus, if $V \in \mathcal{U}$, then according to [46] for any point $x \in V$ we have $\mu_x^V = \varepsilon_x^{E\setminus V}$ (ε_x is the Dirac measure at point x). From this, in particular, we can conclude that a boundary point $x \in \partial V$ is regular if and only if $\varepsilon_x^{E-V} = \varepsilon_x$ – this is a well-known fact of classical potential theory [44], [74], [53].

The set $A \subset E$ is called:

a) *thin at a point* $x \in E$ if $\varepsilon_x^A \neq \varepsilon_x$;
b) *fine open* if the set $E\setminus A$ is thin at all points of set A. The topology generated by the fine open sets is called the fine topology. This topology coincides with the smallest topology in which all hyperharmonic functions are continuous [46].
c) *polar* if $\varepsilon_x^A = 0$ for any $x \in E$.

Any subset of a polar set is polar. In terms of the fine topology a polar set is characterized by the fact that the set of its fine limit points is empty. In terms of the sheaf $\mathcal{S}$, the following characterization applies: there exists a superharmonic function $u \geq 0$ such that $A \subset \{x \in E : u(x) = +\infty\}$;

d) *thin* if it is thin at all points of set E; any countable union of thin sets is called a *semipolar* set. The importance of semipolar sets as exceptional sets in potential theory follows from the *fundamental theorem of convergence* [46: p. 154]. This theorem in classical theory $(\mathbb{R}^n, \mathcal{S}_\Delta)$ was established by Cartan, and in the parabolic theory $(\dot{\mathbb{R}}^n, \mathcal{S}_{\dot{\Delta}})$ by Doob [53].

Theorem. *Let $\{u_i; i \in I\} \subset \mathcal{S}^+(E)$, with u_0 the lower envelop of the family $\{u_i\}$, and let $\underset{+}{u}_0$ be the lower-semicontinuous regularization of the function u_0. Then the set $\{x \in E : \underset{+}{u}_0(x) < u_0(x)\}$ is semipolar.*

We note that the function R_u^A is of a type such that u_0, and therefore also $\{x \in E : \underset{+}{R}_u^A(x) < R_u^A(x)\}$, is a semipolar set.

Every polar set is semipolar. The converse, generally speaking, is not true. For example, in harmonic space $(\dot{\mathbb{R}}^n, \mathcal{S}_{\dot{\Delta}})$ any hypersurface $\{\dot{x} = (x, t) : t = a\}$ is thin and consequently is a semipolar set, but it is not a polar set [53]. In contrast to this example, in classical harmonic space $(\mathbb{R}^n, \mathcal{S}_\Delta)$, the classes of polar and semipolar sets coincide.

In the general case the coincidence of these classes of sets is equivalent to the fulfillment of each of the following properties [46, Chapter 9]:

1) for any locally bounded potential p and any hyperharmonic function $u \geq 0$ we have

$$u \geq p \text{ on } S(p) \ \Rightarrow u \geq p \text{ on } E;$$

2) any locally bounded potential p is continuous on E if its restriction to $S(p)$ is continuous.

The following theorem [2], [5] (see also [39: Chapter 6, 3.14, 3.16]) is key to the probabilistic interpretation of basic concepts of axiomatic potential theory.

Theorem 2.9.1. *For any almost Borel set A and any measure μ with compact support:*

a) $R_u^A(x) = \tilde{P}_A(x, u)$, $\underset{+}{R}_u^A(x) = P_A(x, u)$ $(\forall x \in E, \forall u \in \mathcal{S}^+(E))$,

b) $\mu^A(\Gamma) = P^\mu(x_{T_A} \in \Gamma)$ $(\forall \Gamma \in \mathcal{B})$.

Corollary 2.9.1. *The concepts of a set that is thin at a point, fine open set, polar set, and semipolar set, coincide in the probabilistic and potential senses. In addition, for any set $V \in \mathcal{U}$:*

$$\mu_x^V = P_{E \setminus V}(x, \cdot) \ \ (\forall x \in V).$$

2.10 Duality

Let ν be a measure on E, and X and $\widehat{X}$ be Markov processes on $(E, \mathcal{B})$ with semigroups (P_t) and $(\widehat{P}_t)$ respectively. It is said that processes X and $\widehat{X}$ are dual with respect to ν if:

$$\langle P_t f, g \rangle = \langle f, \widehat{P}_t g \rangle$$

for any functions $f, g \in \mathbb{B}$. Terms relating to process $\widehat{X}$ in this connection are endowed with the prefix "co-", for example, co-potential, co-polar set, etc. In the notation here the symbol "$\wedge$" is used, for example $\widehat{T}_A$, $\widehat{P}_A^\lambda(\cdot, \cdot)$, $\widehat{R}^\lambda$, etc.

The function $r^\lambda(x, y)$ $(x, y \in E, \lambda \geq 0)$ is called a λ-Green's function of processes X and $\widehat{X}$ if:

1) $r^\lambda(x, y)$ is λ-excessive in x and λ-co-excessive in y;
2) for fixed x the function $r^\lambda(x, y)$ is the density of measure $R^\lambda(x, \cdot)$ with respect to measure ν, and for fixed y the function $r^\lambda(x, y)$ is the density of measure $\widehat{R}^\lambda(y, \cdot)$ with respect to ν.

According to [40, Theorem 1.5], a λ-Green's function exists if for each $x \in E$ the measures $R^\lambda(x, \cdot)$ and $\widehat{R}^\lambda(x, \cdot)$ are absolutely continuous with respect to ν. If furthermore the processes X and $\widehat{X}$ are standard, then [40: Theorem 1.6] the fundamental Hunt identity holds:

$$\int r^\lambda(z, y) P_A^\lambda(x, dz) = \int r^\lambda(x, z) \widehat{P}_A^\lambda(y, dz).$$

From this identity many properties can be deduced (see [40: Chapter 6]) which approximate the potential theory of such processes to the classical. Thus, for example, classes of polar (respectively, semipolar) and co-polar (respectively, co-semipolar) sets coincide.

We shall trace the connections of the theory of dual processes with the theory of harmonic spaces. Let $\mathcal{S}$ be a hyperharmonic sheaf such that $(E, \mathcal{H}_{\mathcal{S}})$ is a $\mathcal{P}$-Brelot space. As shown by Hervé [61: Proposition 18.1] (see also [46: Corollary 11.1.3]), with each $y \in E$ it is possible to associate a potential r_y that is harmonic on set $E \backslash \{y\}$. For a given point y any two such potentials, generally speaking, are not proportional [46: Exercise 11.4.9 (g)]. Taking the additional axiom:

(P) (*axiom of proportionality*): any two potentials that are harmonic on a set $E \backslash \{y\}$ are proportional ($\forall y \in E$),

Hervé showed [61: Proposition 18.1] (see also [46: Theorem 11.5.2]) that a function $r(x, y) := r_y(x)$ can be chosen in such a way that these properties are satisfied:

1) r is lower-semicontinuous and continuous off the diagonal;
2) the function $x \to r_y(x)$ is a potential which is harmonic on set $E \backslash \{y\}$ ($\forall y \in E$);
3) for any potential p there exists a unique measure μ such that

$$p = \int r_y \, d\mu(y).$$

Each such function is a candidate for the Green's function (in the above-described probabilistic sense).

For $V \in \mathcal{U}_k$, the function $\underset{+}{R}{}^{E\backslash V}_{r_y}$ is again a potential and therefore can be represented in the form

$$\underset{+}{R}{}^{E\backslash V}_{r_y}(x) = \int r_z(x)\, d\widehat{\varepsilon}^{\,E\backslash V}_y(z).$$

In this connection the measure $\widehat{\varepsilon}^{\,E\backslash V}_y$ is concentrated on ∂V. With the help of the family of measures $\{\widehat{\varepsilon}^{\,E\backslash V}_y\}_{V\in\mathcal{U}_k, y\in V}$, a hyperharmonic sheaf $\widehat{\mathcal{S}}$ can be defined in a natural manner and it has been proved [61: Theorem 29.1] that $(E, \mathcal{H}_{\widehat{\mathcal{S}}})$ is a $\mathcal{P}$-Brelot space; it is called the conjugate space (we note that *apriori* $1 \in \mathcal{S} \cap \widehat{\mathcal{S}}$).

In domain $E \subset \mathbb{R}^n$, if we consider the sheaf $\mathcal{H}_{\mathcal{S}}$ of solutions of second-order partial differential equations of elliptic type with sufficiently regular coefficients, then for a suitable choice of the function $r(x, y)$, the sheaf $\mathcal{H}_{\widehat{\mathcal{S}}}$ will correspond to the classical conjugate equation [61], [62].

For a variation in the choice of function $r(x, y)$, the conjugate harmonic functions are multiplied by some positive continuous function. In [107: Theorem 5.4] (see also [108: Theorem 15]) Taylor showed that if $1 \in \mathcal{S}(E)$ (and for even more general assumptions), then the function $r(x, y)$ can be chosen in such a way that:

1) there exist dual standard processes X and $\widehat{X}$ for which the function $r(x, y)$ is the Green's function;

2) the sheaf $\mathcal{S}$ coincides with the sheaf of superharmonic functions of process X, and sheaf $\widehat{\mathcal{S}}$ coincides with the sheaf of superharmonic functions of process $\widehat{X}$.

Chapter 3
Markov processes and harmonic structures

In this chapter we continue the discussion of the subject of Markov processes and harmonic spaces which we began in Sections 7–10 of Chapter 2. Thus, in the first two sections we shall show that for broad conditions imposed on a Markov process X, harmonic and superharmonic functions with respect to this process satisfy the axioms of harmonic space with a *strong* convergence property (Brelot or Doob). In addition, Secs. 3.1 and 3.2 include a proof of the Harnack inequality (of elliptic (Sec. 3.1) and parabolic (Sec. 3.2) types); the compactness properties of harmonic functions resulting from this inequality are then used in Chapter 5 for the study of weak solutions of infinite-dimensional elliptic equations.

In Secs. 3.3–3.5 we study projective sequences of harmonic spaces and their projective limits. Such sequences arise, for example, in the study of harmonic sheaves generated by space-homogeneous Markov processes on projective Lie groups (see [63: p. 24]) (in particular, on abelian locally compact connected and locally connected groups; see Chapter 4). Examples show that in the general case the limit space does not satisfy all the axioms entering into the definition of a harmonic space (see Chapter 2, Sec. 2.4). Nevertheless, with each such space it is possible to associate a "good" Markov process, and thus a suitable potential theory has developed [18], [23], [104].

3.1 Markov processes and Brelot spaces

In this section we consider a locally compact Hausdorff space E with countable base and a standard Markov process $X = (x_t, \xi, \mathcal{M}_t, P^x)$ in phase space $(E, \mathcal{B})$ with almost surely continuous trajectories. It is assumed also that the topological space E is locally connected and does not have isolated points. For more stringent assumptions about the process X than in the theorem of Bliedtner and Hansen (Theorem 2.8.2), we shall show that:

1) the family of regular sets with respect to sheaf $\mathcal{H}_{\mathcal{S}}$ forms the base of a topology of E;
2) the Brelot convergence property holds, and thus $(E, \mathcal{H}_{\mathcal{S}})$ is a Brelot space.

The potential theory on this space, in the subtlety of its results, is very close to the classical ([44], [74], [46: Chapter 3], [53]).

We assign the standard process X with almost surely continuous trajectories to class $\mathcal{A}$ if the following properties are satisfied:

A.1. There exists a standard process $\widehat{X}$ with almost surely continuous trajectories and dense measure ν on E with respect to which the processes X and $\widehat{X}$ are dual.
A.2. For any function $f \in \mathbb{B}_k$, the functions Rf and $\widehat{R}f$ are bounded and continuous.

Properties A.1 and A.2 imply the absolute continuity of measures $R(x, \cdot)$ and $\widehat{R}(x, \cdot)$ with respect to measure ν for any $x \in E$. Let $r(x, y)$ be the Green's function for processes X and $\widehat{X}$. Our last assumption is:

A.3. The function $r(x, y)$ is continuous in the extended sense on $E \times E$, finite off the diagonal $\Delta = \{(x, y) : x = y\}$ and $r|_\Delta = +\infty$.

The following theorem is the principal one in this section:

Theorem 3.1.1. *If a process X is in a class $\mathcal{A}$, then $\mathcal{S}$ is a hyperharmonic sheaf and $(E, \mathcal{H}_{\mathcal{S}})$ is a $\mathcal{P}$-Brelot space in which these properties are satisfied:*
1) $1 \in \mathcal{S}^+(E)$;
2) *(axiom of proportionality) any two potentials with the given point harmonic support are proportional.*

We shall preface the proof of Theorem 3.1.1 with several auxiliary propositions.

Proposition 3.1.1. *The process X satisfies the conditions of Theorem* 2.8.2 *and therefore $(E, \mathcal{S})$ is a $\mathcal{P}$-harmonic space.*

Proof. Let x_0 be an arbitrary point of E. Since the function $r(x, y)$ is lower-semicontinuous and $r(x_0, x_0) > 0$, then the set $T = \{(x, y) : r(x, y) > 0\}$ is an open neighborhood of the point (x_0, x_0). Let $V, W \in \mathcal{U}_k$ be neighborhoods of x_0 such that $V \times V \subset T$ and $V \backslash \overline{W} \neq \emptyset$. For any point $x \in V \backslash \overline{W}$ we have $\widehat{P}_{E \backslash W}(x, \cdot) = \varepsilon_x(\cdot)$. Now applying the Hunt identity we get

$$0 < r(x_0, x) = \int \widehat{P}_{E \backslash W}(x, dy) r(x_0, y) =$$

$$= \int r(y, x) P_{E \backslash W}(x_0, dy).$$

Therefore the measure $P_{E \backslash W}(x_0, \cdot)$ is non-null and consequently

$$P^{x_0}(T_{E \backslash \{x_0\}} < \xi) > P^{x_0}(T_{E \backslash W} < \xi) = P_{E \backslash W}(x_0, 1) > 0.$$

Thus, condition 1) of Theorem 2.8.2 is satisfied. Condition 2) is satisfied automatically by A.2. Let f be an excessive function. We choose a sequence of compact sets $\{K_n\}$ such that $K_n \subset K_{n+1}$ and $\cup_n K_n = E$ and we set

$$f_n = 1_{K_n}[n(f - nR^n f)] \wedge n.$$

Then by A.2 the functions Rf_n are continuous excessive functions and for $n \uparrow \infty$ they monotonically increase to function f. Thus condition 3) of Theorem 2.8.2 is also satisfied. $\qquad\Box$

We note that the process $\widehat{X}$ is also contained in class $\mathcal{A}$, and therefore according to Proposition 3.1.1, $\widehat{\mathcal{F}}$ is a hyperharmonic sheaf and $(E, \widehat{\mathcal{F}})$ is a $\mathcal{P}$-harmonic space.

Proposition 3.1.2. *The family of regular sets with respect to a sheaf $\mathcal{H}_{\mathcal{F}}$ forms the base of a topology of E.*

Proof. For any point x_0 the function $r_{x_0} = r(\cdot, x_0)$ is superharmonic and $r_{x_0}(x_0) = +\infty$; consequently the set $\{x_0\}$ is polar. Now it remains to use Theorem 6.2.2 from [46], according to which a harmonic space in which every singleton set is polar has a base of a topology consisting of regular sets.

In our situation it is possible to give a simple proof without using the mentioned general result; we shall give it. Let x_0 be an arbitrary point of E. We shall choose an open neighborhood V of this point on which there exists a strictly positive harmonic function h, and we set $V_n = \{x \in V : r_{x_0}(x) > nh(x)\}$. From condition A.3 it follows that V_n is an open subset of V, where for some $\alpha > 0$ and for $n > \alpha$ we have $\overline{V}_n \subset V$. It is clear that the system of sets $\{V_n\}$ forms a base of neighborhoods of point x_0. In addition, it is not difficult to verify that the function $r_{x_0} - nh$ is a barrier for (y, V_n), for any point $y \in \partial V_n$. Therefore the set V_n is regular. $\qquad\Box$

Let μ be a measure on E such that the function

$$R\mu = \int r(\cdot, y)\, d\mu(y)$$

is finite ν-almost everywhere (a.e.) on E. Excessive function $R\mu$ is called the potential of measure μ. Properties of such functions are listed in detail in the monograph of Blumenthal and Getoor [40: Chapter 6]; we note for now only the following property: a finite ν-almost everywhere excessive function p is a potential of some measure μ if and only if the following condition is satisfied:

$$\lim_{V \uparrow E} P_{E\setminus V}(p) = 0 \quad (\nu\text{-almost everywhere})$$

Proposition 3.1.3. *Let supports of measures μ_i $(i = 1, 2)$ be contained in a compact set K. If $R\mu_1(x) \le R\mu_2(x)$ for any $x \in E\setminus K$, then for any co-excessive function f and any neighborhood V of K, the following inequality is satisfied:*

$$\langle \mu_1^{E\setminus V}, f \rangle \le \langle \mu_2^{E\setminus V}, f \rangle \quad (\mu_i^{E\setminus V} = \mu_i \widehat{P}_{E\setminus V}, \; i = 1, 2)$$

Proof. First let $f = \widehat{R}h$ ($h \in \mathbb{B}_K^+$). Applying the Hunt identity and considering that the support of measure $(h\nu)P_{E\setminus V}$ is contained in the set $E\setminus V$, we get

$$
\begin{aligned}
\langle \mu_1^{E\setminus V}, \widehat{R}h \rangle &= \langle \mu_1, \widehat{P}_{E\setminus V}\widehat{R}h \rangle = \\
&= \langle \mu_1, \widehat{R}(h\nu)P_{E\setminus V} \rangle = \langle R\mu_1, (h\nu)P_{E\setminus V} \rangle \le \\
&\le \langle R\mu_2, (h\nu)P_{E\setminus V} \rangle = \langle \mu_2^{E\setminus V}, \widehat{R}h \rangle.
\end{aligned}
$$

In the general case the function f is the limit of an increasing sequence of functions of type $\widehat{R}h$, and therefore the required inequality is established by passing to the limit. $\qquad\square$

Proposition 3.1.4. *Let x_0 be an arbitrary point of E and let $V \in \mathcal{U}_k$ be its co-regular neighborhood (the existence of a base of a topology of E consisting of such neighborhoods follows from Proposition 3.1.2 applied to process $\widehat{X}$). There exists a neighborhood W ($\overline{W} \subset V$) of point x_0 and numbers $\lambda_1, \lambda_2 > 0$ such that for any point $x \in W$*

$$
\lambda_1 P_{E\setminus V}(x_0, \cdot) \le P_{E\setminus V}(x, \cdot) \le \lambda_2 P_{E\setminus V}(x_0, \cdot).
$$

Proof. Let $r_V(x, y)$ ($x, y \in V$) be the Green's function for the part of processes X and $\widehat{X}$ on set V. It is obvious that the function r_V, where

$$
r_V(x, y) = r(x, y) - P_{E\setminus V}(x, r(\cdot, y)),
$$

satisfies condition A.3. Now we shall choose neighborhoods $W, V_0 \in \mathcal{U}_k$ of point x_0 such that $\overline{W} \subset V_0 \subset \overline{V_0} \subset V$ and $\inf_{\overline{V}_0 \times \overline{V}_0} r_V > 0$. Then for some numbers $\lambda_1, \lambda_2 > 0$ and for any $(x, y) \in W \times \partial V_0$ we have:

$$
\lambda_1 r_V(x_0, y) \le r_V(x, y) \le \lambda_2 r_V(x_0, y).
$$

From the co-regularity of V it follows that $r_V(x, y) = 0$ for any $x \in V$ and $y \in \partial V$. Thus, the above-noted inequality is satisfied for any $(x, y) \in W \times \partial(V\setminus\overline{V}_0)$. For any $x \in W$ the function $y \to r_V(x, y)$ is co-harmonic on the set $V\setminus\overline{V}_0$. Therefore the minimum principle implies the satisfaction of the indicated inequality for any $(x, y) \in W \times (V\setminus V_0)$. Now it remains to apply Proposition 3.1.3 for the part of processes X and $\widehat{X}$ on V with excessive function $f = P_{E\setminus V}(g)$ ($g \in \mathbb{B}_K^+$). $\qquad\square$

Corollary 3.1.1. *(elliptic Harnack inequality) Let V be an open connected subset of E. For any compact set $K \subset V$, a number $\lambda = \lambda(K, V) > 0$ exists such that for any non-negative harmonic function h on V the following inequality is satisfied:*

$$
\sup_{x \in K} h(x) \le \lambda \cdot \inf_{x \in K} h(x).
$$

Corollary 3.1.2. *Let $V \in \mathcal{U}$ be a connected set; then:*

1) *any non-negative hyperharmonic function (in particular, a harmonic function) on V is either strictly positive or identically zero;*
2) *any non-negative hyperharmonic function on V is either finite on a dense set or identically equals $+\infty$;*
3) *for any point $x \in V$, the support of measure $P_{E\backslash V}(x, \cdot)$ coincides with ∂V.*

We omit the standard proof of these properties.

Proof of Theorem 3.1.1. According to Proposition 3.1.1, $\mathcal{S}$ is a hyperharmonic sheaf and $(E, \mathcal{S})$ is a $\mathcal{P}$-harmonic space. By Proposition 3.1.2, a family of regular sets forms the base of a topology of E, therefore $(E, \mathcal{H}_{\mathcal{S}})$ is a $\mathcal{P}$-Bauer space. From the Harnack inequality (Corollary 3.1.1), it follows that the Brelot convergence property is satisfied in $(E, \mathcal{H}_{\mathcal{S}})$, and thus it is proved that $(E, \mathcal{H}_{\mathcal{S}})$ is a $\mathcal{P}$-Brelot space. The inclusion $1 \in \mathcal{S}^+(E)$ is obvious. We shall show that the proportionality axiom is satisfied. Let p be a potential with harmonic support $\{x_0\}$. We shall show that the functions p and $r_{x_0} = r(\cdot, x_0)$ are proportional. We choose a sequence $\{V_n\}$ of sets of $\mathcal{U}_k$ such that $V_n \subset V_{n+1}$ and $\cup_n V_n = E$. The sequence of functions $\{P_{E\backslash V_n}(p)\}$ is monotonically decreasing; let $\overline{p} = \inf_n P_{E\backslash V_n}(p)$. The function $\overline{p}$ is non-negative, harmonic on E, and is dominated by the potential p, and consequently $\overline{p} \equiv 0$. But then, as was noted before Proposition 3.1.3, there exists a measure μ such that $p = R\mu$. Now let the set $V \in \mathcal{U}_k$ be such that $\overline{V} \cap \{x_0\} = \emptyset$. Considering the harmonicity of p on V and applying the Hunt identity, we get

$$p = P_{E\backslash V}(p) = R\mu\widehat{P}_{E\backslash V}.$$

From this it follows that $\mu = \mu\widehat{P}_{E\backslash V}$ and, in particular, that $\mathrm{supp}\mu \cap V = \emptyset$. In view of the arbitrariness of the choice of $V \subset E\backslash\{x_0\}$, we conclude that $\mathrm{supp}\mu = \{x_0\}$. Consequently for some number $c > 0$, we have $\mu = c \cdot \varepsilon_{x_0}$ and thus $p = c \cdot r_{x_0}$. $\square$

Remark 3.1.1. The dual process $\widehat{X}$ is also in class $\mathcal{A}$, and therefore $(E, \mathcal{H}_{\widehat{\mathcal{S}}})$ is a $\mathcal{P}$-Brelot space with properties 1) and 2). In this connection harmonic spaces $(E, \mathcal{H}_{\mathcal{S}})$ and $(E, \mathcal{H}_{\widehat{\mathcal{S}}})$ are conjugate (with respect to a given Green's function r; see Chapter 2, Sec. 2.10).

In the case when E is a Euclidean space or a domain of it, conditions A.1–A.3 are satisfied for a wide class of diffusion processes ([54: Chapter 13, § 5]). In this connection, functions of $\mathcal{H}_{\mathcal{S}}$ are solutions of a corresponding second-order partial differential equation of elliptic type. Functions from $\mathcal{H}_{\widehat{\mathcal{S}}}$ are solutions of the classical conjugate equation (see also [61], [62]). In the following chapters we shall consider infinite-dimensional elliptic equations on the group $\mathbb{R}^n \times \mathbb{T}^\infty$.

3.2 Markov processes and Bauer spaces

In this section we assume that topological space E is locally connected and has no isolated points. We shall consider a class $\mathcal{B}$ of Markov processes on E which is defined in the following manner. A standard process X with almost surely continuous trajectories is contained in a class $\mathcal{B}$ if the following conditions are satisfied:

B.1. There exists a standard process $\widehat{X}$ with almost surely continuous trajectories and a dense measure ν on E with respect to which the processes X and $\widehat{X}$ are dual.

B.2. There exists a function $p(t; x, y)$ $(t > 0, x, y \in E)$ such that:

(1) the function $p(t; x, y)$ is continuous with respect to $(t; x, y)$ for $t > 0$;

(2) $p(t; x, y) \to 0$ when $t \downarrow 0$ uniformly with respect to (x, y) on every compact set $K \subset (E \times E)\backslash\Delta$;

(3) $\lim_{t \downarrow 0} p(t; x, x) = +\infty$ $(\forall x \in E)$;

(4) $P_t f(x) = \int_E p(t; x, y) f(y)\, d\nu(y)$, $\widehat{P}_t f(x) = \int_E p(t; y, x) f(y)\, d\nu(y)$ $(\forall f \in \mathbb{B})$.

Let Z be the Markov process on the line $\mathbb{R}^1$ consisting of uniform motion to the left with a speed equal to one. We shall form a new process $\dot{X} = X \times Z$ in phase space $(\dot{E}, \dot{\mathcal{B}})$: $\dot{E} = E \times \mathbb{R}^1$, $\dot{\mathcal{B}} = \mathcal{B}(\dot{E})$. The Markov semigroup $(\dot{P}_t)$ of process $\dot{X}$ has the form

$$\dot{P}_\tau f(\dot{x}) = P_\tau f(\cdot, t - \tau)(x), \quad \dot{x} = (x, t) \in \dot{E}.$$

Let $\widehat{Z}$ be the Markov process on the line $\mathbb{R}^1$ consisting of uniform motion to the right with speed equal to one and $\dot{\widehat{X}} = \widehat{X} \times \widehat{Z}$ a Markov process in phase space $(\dot{E}, \dot{\mathcal{B}})$. It is not difficult to see that the processes $\dot{X}$ and $\dot{\widehat{X}}$ are dual with respect to the measure $\dot{\nu} = \nu \otimes ds$ (ds is the Lebesgue measure on the line) and have Green's function $\dot{r}(\dot{x}, \dot{y})$, $\dot{x} = (x, t)$, $\dot{y} = (y, s)$, which has the form

$$\dot{r}(\dot{x}, \dot{y}) = \begin{cases} p(t - s; x, y), & t > s \\ 0, & t \leq s. \end{cases}$$

The function $\dot{r}(\dot{x}, \dot{y})$ is lower-semicontinuous and continuous on the set $(\dot{E} \times \dot{E})\backslash\dot{\Delta}$. However, in distinction from the Green's function for processes from class $\mathcal{A}$, its continuity at points of diagonal $\dot{\Delta}$ is broken. This distinction plays an essential role from the point of view of potential theory. In fact, sheaf $\mathcal{H}_{\dot{\varphi}}$ of harmonic functions of process $\dot{X}$ *will have the Doob convergence property and not have the Brelot convergence property*. The fact that the Brelot convergence property is not satisfied in this situation can now be easily explained. Thus, for any point $\dot{y} \in \dot{E}$, the function $\dot{r}_{\dot{y}}(\dot{x}) = \dot{r}(\dot{x}, \dot{y})$ is superharmonic on set $\dot{E}$ and harmonic on set $\dot{E}\backslash\{\dot{y}\}$. For any open connected set $\dot{V}$ which does not contain the point $\dot{y} = (y, s)$ and has non-empty intersection with each of the semispaces $\dot{E}_s^+ = \{(x, t) : t > s\}$ and $\dot{E}_s^- = \{(x, t) : t < s\}$, the sequence $u_n := n \dot{r}_{\dot{y}}$ of harmonic functions on $\dot{V}$ increases monotonically and goes to zero on the set $\dot{V}^- = \dot{V} \cap \dot{E}_s^-$ and to infinity at all points of set $\dot{V}^+ = \dot{V} \cap \dot{E}_s^+$ at which $\dot{r}_{\dot{y}} > 0$; it is obvious that the set of such points is a non-empty subset of V^+. Thus the function $u_\infty = \sup_{n>0} u_n$ is finite on $\dot{V}^-$ and infinite on a non-empty subset of $\dot{V}^+$; thus it is shown that *the Brelot convergence property is not satisfied.*

In the case when E is a Euclidean space or a domain of it, conditions B.1 and B.2 are satisfied for a wide class of diffusion processes [54: Chapter 13, § 5]. In this connection the functions from $\mathcal{H}_{\mathcal{G}}$ are the solutions of a second-order partial differential equation of parabolic type $(\mathcal{L} - \partial_t)u = 0$, in which $\mathcal{L}$ is the elliptic differential operator corresponding to process X. The functions from $\mathcal{H}_{\widehat{\mathcal{G}}}$ are the solutions of conjugate equation $(\mathcal{L}^* + \partial_t)u = 0$. We note also that in the classical case X is a Wiener process, $\dot{X}$ is a space-time Wiener process, and $\mathcal{H}_{\mathcal{G}}$ is the sheaf of solutions of the heat equation ([53], [66: 7.14]).

The principal result of this section is contained in the following theorem.

Theorem 3.2.1. *If process X is in class $\mathcal{B}$, then $\dot{\mathcal{G}}$ is a hyperharmonic sheaf and $(\dot{E}, \mathcal{H}_{\mathcal{G}})$ is a $\mathcal{P}$-Bauer space in which the Doob convergence property holds.*

The proof of this theorem is carried out by the same plan as the proof of the similar Theorem 3.1.1. However in some details there is an essential difference connected first of all with the fact that the "parabolic" Green's function $\dot{r}$ goes to zero on diagonal $\dot{\Delta}$.

Proposition 3.2.1. *The process $\dot{X}$ satisfies the conditions of Theorem 2.8.2 and therefore $(\dot{E}, \dot{\mathcal{G}})$ is a $\mathcal{P}$-harmonic space.*

Proof. Fulfillment of the condition $P^{\dot{x}}(T_{\dot{E}\setminus\{\dot{x}\}} < \dot{\xi}) > 0$ for any point $\dot{x} \in \dot{E}$ is obvious. For any compact set $\dot{K} \subset \dot{E}$ of type $K \times I$, where K is a compact set in E, and I is a compact interval on the line, we have

$$\dot{R}1_{\dot{K}}(\dot{x}) = \int_0^\infty \dot{P}_t 1_{\dot{K}}(\dot{x})\, dt =$$

$$= \int_0^\infty P_t 1_K(x) 1_I(\tau - t)\, dt \leq |I| < +\infty,$$

where $\dot{x} = (x, \tau)$ is an arbitrary point of $\dot{E}$; thus the process $\dot{X}$ also satisfies the second condition of Theorem 2.8.2. For any function $f \in \mathbb{B}_{\dot{K}}^+(\dot{E})$, the support of which is contained in a compact set of the form $K \times I$, we write the equality

$$\dot{R}f(\dot{x}) = \int_0^\infty P_t f(\cdot, \tau - t)(x)\, dt = \int_0^\delta + \int_\delta^\infty := \dot{R}_1 + \dot{R}_2.$$

We have: $\dot{R}_1 \leq \delta \|f\|$, and for $\dot{R}_2$ we write the equality:

$$\dot{R}_2 = \int_{-\infty}^{\tau - \delta} P_{\tau - s} f(\cdot, s)(x)\, ds = \int_{\dot{K}_\tau} p(\tau - s; x, y) f(\dot{y})\, d\dot{v}(\dot{y}),$$

where $\dot{y} = (y, s)$, $\dot{v} = v \otimes ds$, $\dot{K}_\tau = K \times (I \cap\ ]-\infty, \tau - \delta])$.

From this equality and from the continuity of the function $p(t; x, y)$ on the set $\{(t; x, y) : t > \delta\}$, we get the continuity of both functions $\dot{R}_2$ and $\dot{R}f$. Thus, in turn, the last condition of Theorem 2.8.2 is satisfied. $\square$

The process $\widehat{X}$ is also in class $\mathcal{B}$. The same arguments as in the proof of Proposition 3.1.1, but applied to the process $\widehat{\dot{X}} = \widehat{X} \times \widehat{Z}$, show that the sheaf $\widehat{\mathcal{G}}$ of locally superharmonic functions corresponding to this process is a hyperharmonic sheaf and $(\dot{E}, \widehat{\mathcal{G}})$ is a $\mathcal{P}$-harmonic space.

Proposition 3.2.2. *The family of regular sets with respect to a sheaf $\mathcal{H}_{\mathcal{G}}$ forms the base of a topology of space $\dot{E}$.*

Proof. Here we intend to use the general result [46: Theorem 6.2.2] according to which every harmonic space in which all singleton sets are polar has a base of a topology consisting of regular sets. We shall show that in the space $(\dot{E}, \dot{\mathcal{G}})$ every one-point set is polar. Let $\dot{x}_0 = (x_0, t_0)$ be an arbitrary point of $\dot{E}$. Since $p(t; x_0, x_0) \to +\infty$ for $t \downarrow 0$, it is possible to choose a sequence $\{t_k\}$ such that $t_k \downarrow 0$ and

$$\sum_{k=1}^{\infty} 1/p(t_k; x_0, x_0) < +\infty.$$

We set $\dot{x}_k = (x_0, t_0 - t_k), c_k = 1/p(t_k; x_0, x_0)$ and we consider the measure $\mu = \sum_{k=1}^{\infty} c_k \varepsilon_{\dot{x}_k}$. We have $\mu(1) = \sum_{k=1}^{\infty} c_k < \infty$, and in addition for any function $\varphi \in \mathbb{C}_k^+(\dot{E})$ we have

$$\langle \varphi, \dot{R}\mu \rangle = \langle \widehat{\dot{R}}\varphi, \mu \rangle \leq \|\widehat{\dot{R}}\varphi\| \mu(1) < \infty.$$

Thus the hyperharmonic function $\dot{R}\mu$ is locally summable with respect to measure $\dot{\nu}$ and, in particular, is finite on a dense subset in $\dot{E}$; we shall calculate its value at a point $\dot{x}_0$:

$$\dot{R}\mu(\dot{x}_0) = \sum_{k=1}^{\infty} c_k \dot{r}(\dot{x}_0, \dot{x}_k) = \sum_{k=1}^{\infty} 1/p(t_k; x_0, x_0) \cdot p(t_k; x_0, x_0) = \infty.$$

Consequently $\{\dot{x}_0\}$ is a polar set. In order to finish the proof it remains to note that the point $\dot{x}_0$ was chosen arbitrarily. $\qquad \square$

Let W be an open subset of $\dot{E}$ and let $\dot{r}_W(\dot{x}, \dot{y})$ be the Green's function for the part of processes $\dot{X}$ and $\widehat{\dot{X}}$ on W:

$$\dot{r}_W(\dot{x}, \dot{y}) = \dot{r}(\dot{x}, \dot{y}) - P_{\dot{E} \backslash W}(\dot{x}, \dot{r}(\cdot, \dot{y})).$$

The function $\dot{r}_W(\dot{x}, \dot{y})$ is lower-semicontinuous on $W \times W$, continuous on $(W \times W) \backslash \dot{\Delta}$, and for $t \leq s$ ($\dot{x} = (x, t), \dot{y} = (y, s)$) it vanishes. For a point $\dot{a} \in W$ and a number $\gamma > 0$ we set

$$B_{W, \dot{a}, \gamma} = \{\dot{y} \in W : \dot{r}_W(\dot{a}, \dot{y}) > \gamma\}.$$

From the lower-semicontinuity of $\dot{r}_W$ it follows that $B_{W,\dot{a},\gamma}$ is an open subset of W, and if W is co-regular (according to Proposition 3.2.2 applied to process $\widehat{\dot{X}}$, such sets form the base of a topology of $\dot{E}$), then $\overline{B}_{W,\dot{a},\gamma} \subset W$.

Proposition 3.2.3. *Let $\dot{a}$ be an arbitrary point of $\dot{E}$ and let W be its co-regular neighborhood. For any compact set $K \subset B_{W,\dot{a},\gamma}$ there exists a neighborhood V_K of point $\dot{a}$ and a number $\lambda_K > 0$ such that:*
1) $\exists \delta \in \mathbb{R}^1 : K \subset E \times \;]-\infty, \delta[\;, \; \overline{V}_K \subset E \times \;]\delta, +\infty[\;;$
2) for any non-negative function $h \in \mathcal{H}_{\dot{\mathcal{G}}}(W)$ we have

$$\sup_K h \leq \lambda_K \inf_{V_K} h.$$

Proof. For any number β and any subset $V \subset \dot{E}$, we set

$$V_\beta^+ = \{\dot{x} = (x,t) \in V : t > \beta\}, \quad V_\beta^- = \{\dot{x} = (x,t) \in V : t < \beta\}.$$

Now let K be a compact subset of $B_{W,\dot{a},\gamma}$. Since $B_{W,\dot{a},\gamma} \subset \dot{E}_\alpha^-$, $\dot{a} = (a,\alpha)$, there exists a number $\delta < \alpha$ such that $K \subset \dot{E}_\delta^-$. From the fact that the function $\dot{r}_W$ is continuous on $(W \times W) \backslash \dot{\Delta}$, it follows that for $\dot{y} \to \dot{a}$, we have $\dot{r}_W(\dot{y}, \dot{z}) \to \dot{r}_W(\dot{a}, \dot{z})$ uniformly on every compact set not containing the point $\dot{a}$. Thus we deduce that there exists a neighborhood V_K of point $\dot{a}$ such that $\overline{V}_K \subset E_\delta^+$ and in addition

$$m := \inf\{\dot{r}_W(\dot{y}, \dot{z}) : \dot{y} \in \overline{V}_K, \dot{z} \in \overline{B}_{W,\dot{a},\gamma} \cap \overline{W}_\delta^-\} > \frac{\gamma}{2} > 0.$$

Further, from the continuity of $\dot{r}_W$ on $(W \times W) \backslash \dot{\Delta}$ it follows also that:

$$M := \sup\{\dot{r}_W(\dot{x}, \dot{z}) : \dot{x} \in K, \dot{z} \in \overline{W} \backslash B_{W,\dot{a},\gamma}\} < +\infty.$$

We set $\lambda_K = \frac{M}{m}$ and show that the inequality

$$\dot{r}_W(\dot{x}, \dot{z}) \leq \lambda_K \dot{r}_W(\dot{y}, \dot{z}),$$

$$\dot{x} \in K, \; \dot{y} \in V_K, \; \dot{z} \in W \backslash (\overline{V}_K \cup \overline{B}_{W,\dot{a},\gamma})$$

is satisfied. We fix $\dot{x} \in K$ and $\dot{y} \in V_K$ arbitrarily and note that the right- and left-hand sides of the inequality are co-harmonic functions on $W \backslash (\overline{V}_K \cup \overline{B}_{W,\dot{a},\gamma})$, vanishing on ∂W. Further, the inequality under investigation is trivially fulfilled for $\dot{z} \in W_\delta^+$, since for such $\dot{z}$ the left-hand side of the inequality vanishes. And finally, for $\dot{z} \in \partial B_{W,\dot{a},\gamma} \cap W_\delta^-$ this inequality is satisfied by the choice of number λ_K. Thus the inequality is satisfied on the set $\partial W \cup \partial V_K \cup \partial B_{W,\dot{a},\gamma}$, which obviously contains the boundary of $W \backslash (\overline{V}_K \cup \overline{B}_{W,\dot{a},\gamma})$. From the minimum principle it now follows that the inequality is satisfied everywhere on the set $W \backslash (\overline{V}_K \cup \overline{B}_{W,\dot{a},\gamma})$.

From here on, the proof of the proposition concludes like the proof of the similar Proposition 3.1.4. $\qquad\qquad\square$

Proof of Theorem 3.2.1. According to Proposition 3.2.1, $\mathcal{S}$ is a hyperharmonic sheaf and $(\dot{E}, \mathcal{S})$ is a $\mathcal{P}$-harmonic space. By Proposition 3.2.2, the family of regular sets forms the base of a topology of $\dot{E}$, and therefore $(\dot{E}, \mathcal{H}_{\mathcal{S}})$ is a $\mathcal{P}$-Bauer space. We shall derive the Doob convergence property from Proposition 3.2.3, in which the local form of the "parabolic" Harnack inequality is given. Let $V \in \mathcal{U}$ and $\{h_n\}$ be an increasing sequence of harmonic functions on V such that $\sup_n h_n < \infty$ on a dense subset of V. Also let $\dot{x}_0 = (x_0, t_0)$ be a fixed point of V and let W be a sufficiently small co-regular neighborhood of this point. Since $\dot{r}_W(\dot{x}_0, \dot{x}) \to \infty$ for $\dot{x} = (x_0, t) \to \dot{x}_0$ $(t < t_0)$, then the set $B_{W, \dot{x}_0, \gamma}$ contains some "segment" $\{x_0\} \times \ ]\delta, t_0[$. Let the number $t_1 \in \ ]\delta, t_0[$ be such that for $\Delta = t_0 - t_1$ the set $\tau_\Delta V = \{\dot{x} = (x, t) : \tau_\Delta \dot{x} = (x, t+\Delta) \in V\}$ contains W. Then the functions $h_n^\Delta(\dot{x}) = h_n(\tau_\Delta \dot{x})$ are harmonic on W, the sequence $\{h_n^\Delta\}$ is monotonically increasing, and $\sup_n h_n^\Delta < \infty$ on a dense subset of $\tau_\Delta V$ and in particular, on a dense subset of W. From Proposition 3.2.3 it follows that the sequence $\{h_n^\Delta\}$ is bounded on every compact subset of $B_{W, \dot{x}_0, \gamma}$. In particular, the sequence $\{h_n^\Delta\}$ is bounded on some neighborhood of point $\dot{x}_1 = (x_0, t_1)$, and thus the sequence $\{h_n\}$ is bounded on some neighborhood of point $\tau_\Delta \dot{x}_1 = \dot{x}_0$. In view of the arbitrariness of the choice of point $\dot{x}_0$, we conclude that the sequence $\{h_n\}$ is locally bounded and consequently $\sup_n h_n$ is a harmonic function. $\square$

Corollary 3.2.1. *Let X be a Markov process on E with almost surely continuous trajectories and let $\mathcal{S}$ be the sheaf of its locally superharmonic functions. If the process X is in class $\mathcal{B}$, then $\mathcal{S}$ is a hyperharmonic sheaf and $(E, \mathcal{H}_{\mathcal{S}})$ is a Brelot space.*

Proof. From the existence of transition density $p(t; x, y)$ of the process X, it follows [54: Chapter 5, Lemma 5.3] that there are no absorbent points in E, that is, points x such that

$$P^x(T_{E \setminus \{x\}} = +\infty) = 1,$$

thus, in turn, it follows [54: Chapter 5, Lemma 5.5] that there exists a neighborhood $V \in \mathcal{U}_K$ of point x such that

$$\sup_{z \in E} M_z(T_{E \setminus V}) < \infty.$$

Therefore λ-potential operator R_V^λ for process X_V , the part of process X on V, is defined also for $\lambda = 0$ (which, generally speaking, cannot be said of the analogous operator for initial process X); we note in passing that process X_V is in class $\mathcal{B}$. Further, for any open set $W : W \subset V$, the families of locally superharmonic functions on W for processes X and X_V respectively coincide. Considering the above-said and the fact that the assertion we have proved has local character, it is possible, without loss of generality, to assume that process X satisfies the condition:

$$\sup_{z \in E} R1(z) < \infty.$$

We shall use Theorem 2.8.2 first. Conditions 1) and 2) of this theorem are trivially satisfied; we shall verify condition 3). Let f be an excessive function with respect to X and let π be the projection of $\dot{E}$ onto E. It is obvious that the function $\dot{f} = f \cdot \pi$ is excessive with respect to process $\dot{X}$ and consequently is lower-semicontinuous, and thus function f is also lower-semicontinuous. Let g be a function in $\mathbb{C}_K^+$ such that $g \leq f$ and $\dot{R}_{\dot{g}}$ is the reduction (with respect to sheaf $\dot{\mathcal{S}}$) of function $\dot{g} = g \cdot \pi$. According to [46: Proposition 2.2.3], the function $\dot{R}_{\dot{g}}$ is continuous and excessive with respect to process $\dot{X}$; in addition,

$$\dot{g} \leq \dot{R}_{\dot{g}} \leq \dot{f}.$$

According to [39: V.5.11, 8.1],

$$\dot{R}_{\dot{g}} = R_g \cdot \pi,$$

where R_g is the reduction of g determined by $\mathcal{S}$. It is clear that function R_g is continuous, excessive with respect to process X, and increases when g does; we have:

$$g \leq R_g \leq f.$$

Passing to sup on the left-hand side of the inequality over all such functions g, we obtain $R_g \uparrow f$; consequently condition 3) of Theorem 2.8.2 is satisfied. According to this theorem, $\mathcal{S}$ is a hyperharmonic sheaf and $(E, \mathcal{S})$ is a $\mathcal{P}$-harmonic space.

We shall show that the Brelot convergence axiom is satisfied in $(E, \mathcal{S})$. Let x_0 be an arbitrary point of E, and let t_0 be any number and W some co-regular neighborhood of point $\dot{x}_0 = (x_0, t_0)$. We shall fix a number $\gamma > 0$ and consider the set $B_{W, \dot{x}_0, \gamma}$. As already noted, this set contains some "interval" $\{x_0\} \times \ ]\beta, t_0[$. We choose a number t_1 and neighborhood $W_0 \in \mathcal{U}_K$ of point x_0 such that the compact set $K = \overline{W}_0 \times \{t_1\}$ is contained in set $B_{W, \dot{x}_0, \gamma}$. In correspondence with Proposition 3.2.3, with respect to the given compact set K we shall choose a neighborhood V_K of point $\dot{x}_0$ and a number $\lambda_K > 0$ such that $\forall h \in \mathcal{H}_{\dot{\mathcal{S}}}^+(W)$

$$\sup_{K} \dot{h} \leq \lambda_K \inf_{V_K} \dot{h}$$

Shrinking neighborhoods V_K and W_0 if necessary, we may assume that $V_K = W_0 \times I$, where I is a compact interval containing point t_0. Now applying the above-noted inequality to harmonic functions $\dot{h}$ of type $h \cdot \pi$, we arrive at the local Harnack inequality for harmonic functions of process X: for any point x_0 and any neighborhood $V \in \mathcal{U}_K$ of it, there exists its neighborhood $W_0 \in \mathcal{U}_K : \overline{W}_0 \subset V$ and a number $\lambda = \lambda_{V,x_0} > 0$ such that for any function $h \in \mathcal{H}_{\mathcal{G}}^+(V)$ the inequality

$$\sup_{\overline{W}_0} h \leq \lambda \cdot \inf_{\overline{W}_0} h$$

is satisfied. From this inequality it is easy to deduce the Brelot convergence property (and at the same time the global Harnack inequality: see Corollary 3.1.1, as well as [97]).

We shall show that the *elliptic property* is satisfied in $(E, \mathcal{G})$: any nonnegative hyperharmonic function on a connected open set is either strictly positive or identically zero. According to [46: Proposition 3.1.4 e)$\Rightarrow$d)] from this property it follows that in $(E, \mathcal{G})$ the family of regular sets forms the base of a topology of E. Thus, verifying the *elliptic property*, we arrive at the fact that $(E, \mathcal{H}_{\mathcal{G}})$ is a Bauer space with the Brelot convergence property and consequently, $(E, \mathcal{H}_{\mathcal{G}})$ is a Brelot space.

Let V be a connected open subset of E and let function $f \in \mathcal{G}^+(V)$ be such that $f(x_0) = 0$ for some point $x_0 \in V$; we shall show that $f \equiv 0$. If $f \in \mathcal{H}_{\mathcal{G}}^+(V)$, then the identity $f \equiv 0$ immediately follows from the Harnack inequality. In the general case we set $f_n = \min(f, n)$. The function f_n is superharmonic and if u_n is its nonnegative harmonic minorant, then $u_n(x_0) = 0$ and consequently $u_n \equiv 0$. Therefore f_n is a potential. Let μ be a measure on V such that $f_n = R_V \mu$. Since $f_n(x_0) = 0$, then either $\mu = 0$ or there exists a point $y_0 \in V$ such that $r_V(x_0, y_0) = 0$; we note that $y_0 \neq x_0$ since

$$r_V(x_0, x_0) = \int_0^\infty p_V(t; x_0, x_0)\, dt > 0,$$

where $p_V(t; x, y)$ is the transition density of process X_V. In the first case $f_n \equiv 0$; therefore we shall consider the second case. We set $A = \{y \in V : r_V(x_0, y) = 0\}$; according to the assumption, $A \neq \emptyset$. The function $r_V(x_0, y)$ is co-superharmonic and therefore is lower-semicontinuous, and consequently A is a closed subset of V. Further, the function $r_V(x_0, y)$ is co-harmonic on the set $V \backslash \{x_0\}$. Since $A \subset V \backslash \{x_0\}$, A is an open subset of V. From the connectedness of V and the fact that $A \neq V$, it now follows that $A = \emptyset$; contradiction. Thus, $f_n \equiv 0$ and therefore $f \equiv 0$. $\quad\square$

Remark 3.2.1. The harmonic space $(\dot{E}, \widehat{\mathcal{G}})$ corresponding to the dual process $\widehat{X} = \widehat{X} \times \widehat{Z}$ is conjugate to space $(\dot{E}, \mathcal{H}_{\dot{\mathcal{G}}})$ with respect to Green's function $\dot{r}$ (Chapter 2, Section 2.10). Since the process $\widehat{X}$ is also in class $\mathcal{B}$, the arguments which we carried out for process $\dot{X}$, with obvious changes, are applicable also to process $\widehat{X}$. Consequently, $(\dot{E}, \mathcal{H}_{\widehat{\mathcal{G}}})$ is a $\mathcal{P}$-Bauer space with the Doob convergence axiom. Thus in this case the conjugate space has the same properties as the original.

The inequality from Proposition 3.2.3 is the local form of the "parabolic" Harnack inequality. The classical form of this inequality (see for example [73], [46]) is given in Proposition 3.2.4.

Following [46: 6.1], we shall call a closed set $A \subset \dot{E}$ *absorbent* if a function equal to zero on A and to $+\infty$ on $\dot{E} \backslash A$ is hyperharmonic. According to [46: Proposition 6.1.1] any closed set which, together with every point $\dot{x}$ and any neighborhood V of it, contains a support of harmonic measure $P_{\dot{E}\backslash V}(\dot{x}, \cdot)$, is absorbent. In particular, every closed semispace $\overline{\dot{E}}_{\tau}^{-} = \{\dot{x} = (x, t) : t \leq \tau\}$ is an absorbent set.[1]

Proposition 3.2.4. *Let $\mathcal{F}$ be the base of a topology of space E consisting of connected sets and $\mathcal{I}$ the family of open finite intervals. The following assertions are equivalent:*

1) *for any set $V \in \mathcal{F}$ the transition density $p_V(t; x, y)$ of the part of process X on set V is strictly positive;*
2) *for any set $\dot{V} = V \times I$ ($V \in \mathcal{F}, I \in \mathcal{I}$) and compact sets $K_- \subset \dot{V}_{\tau}^{-}, K_+ \subset \dot{V}_{\tau}^{+}$ ($\tau \in I$) there exists a number $\lambda > 0$ such that for any function $h \in \mathcal{H}_{\dot{\mathcal{I}}}^{+}(\dot{V})$ the inequality*

$$\sup_{K_-} h \leq \lambda \inf_{K_+} h$$

is satisfied;
3) *for any set $\dot{V} = V \times I$ ($V \in \mathcal{F}, I \in \mathcal{I}$) with regular base[2] V and any point $\dot{x} = (x, t) \in \dot{V}$, the support of harmonic measure $P_{\dot{E}\backslash\dot{V}}(\dot{x}, \cdot)$ coincides with the set $\partial\dot{V} \cap \overline{\dot{E}_t^{-}}$;*
4) *for any set $\dot{V} = V \times I$ ($V \in \mathcal{F}, I \in \mathcal{I}$), sets of type $\dot{V} \cap \overline{\dot{E}_t^{-}}$ and only those are absorbent subsets of the harmonic space $(\dot{V}, \dot{\mathcal{I}})$.*

In this connection, if the equivalent assertions 1)–4) are true for some base $\mathcal{F}_0$, then they are true also for any base of $\mathcal{F}$ consisting of connected sets.

Before proceeding with the proof of the proposition, we note that if $E = \mathbb{R}^n$ and X is a Wiener process, then the truth of 1) is easy to verify on the parallelepipeds $V = I_1 \times \ldots \times I_n$. In fact, in this case

$$p_V(t; x, y) = \prod_{k=1}^{n} p_{I_k}(t; x_k, y_k) \quad (x = (x_k)_1^n, \, y = (y_k)_1^n),$$

where $p_{I.}(t; x., y.)$ is the transition density of the part of a Wiener process on interval $I..$ It is known [53: 2.VII.II] that the function $p_I(t; x, y)$ is expressed by the function $n_t(x) = (2\pi t)^{-\frac{1}{2}} \exp(-x^2/2t)$ according to the formula

$$p_I(t; x, y) = \sum_{k=-\infty}^{\infty} [n_t(2kc - x + y) - n_t(2kc + 2a - x - y)],$$

where $I = \,]a, b[\,, \, c = b - a$. Verification of the strict positivity of function p_I can now be carried out easily [46: 3.3, Lemma 2c)].

---

1   Here and below we use the notation $\dot{V}_{\tau}^{+}, \dot{V}_{\tau}^{-}$ introduced in the proof of Proposition 3.2.3.
2   I.e., the set $V \subset E$ is regular with respect to sheaf $\mathcal{I}$ generated by process X.

Proof of Proposition 3.2.4. 1)$\Rightarrow$2): Here we apply the same method as in the proof of Proposition 3.2.3. We fix a set $\dot{V} = V \times I$, a number $\gamma \in I$, and compact sets $K_- \subset \dot{V}_\gamma^-$ and $K_+ \subset \dot{V}_\gamma^+$. We note further that for points $\dot{x}, \dot{y} \in \dot{V}$ the equality

$$\dot{r}_{\dot{V}}(\dot{x}, \dot{y}) = p_V(t - s; x, y) \quad (\dot{x} = (x, t), \dot{y} = (y, x))$$

holds. In correspondence with Property 1) we choose the set $\dot{W} = W \times I$ and numbers $\varepsilon, \tau > 0$ such that $K_- \cup K_+ \subset \dot{W} \subset \overline{\dot{W}} \subset \dot{V}$ and

$$r_{\dot{V}}|_{K_- \times \partial \dot{W}_\gamma^-} \leq \tau, \; \dot{r}_{\dot{V}}|_{K_+ \times \partial \dot{W}_\gamma^-} \geq \varepsilon.$$

Setting $\lambda = \tau/\varepsilon$, as in the proof of Proposition 3.2.3, we arrive at the fact that the inequality

$$\dot{r}_{\dot{V}}(\dot{x}, \dot{z}) \leq \lambda \dot{r}_{\dot{V}}(\dot{y}, \dot{z})$$

holds for any $\dot{x} \in K_-$, $\dot{y} \in K_+$, and $\dot{z} \in \partial \dot{W}$. The truth of this inequality for $\dot{z} \in \dot{V} \backslash \overline{\dot{W}}$ follows from the following form of the minimum principal [39: III. 6.6]: let $\mathcal{V}$ be an open subset of harmonic space $(\mathcal{E}, \mathcal{S})$, $u \in \mathcal{S}(\mathcal{V})$, and p a potential such that

a) $\forall z \in \partial \mathcal{V}$, we have $\liminf_{x \to z} u(x) \geq 0$,

b) $\forall z \in \mathcal{V}$, we have $u(z) \geq -p(z)$,

then $u(z) \geq 0$ $(\forall z \in \mathcal{V})$. In this case one should set:

$$\mathcal{E} := \dot{V}, \quad \mathcal{V} := \dot{V} \backslash \overline{\dot{W}}, \quad \mathcal{S} := \widehat{\dot{\mathcal{S}}},$$

$$u(\dot{z}) := \lambda \dot{r}_{\dot{V}}(\dot{y}, \dot{z}) - \dot{r}_{\dot{V}}(\dot{x}, \dot{z}), \; p(\dot{z}) := \dot{r}_{\dot{V}}(\dot{x}, \dot{z}).$$

Now applying Proposition 3.1.3 with function $h \in \mathcal{H}_{\dot{\mathcal{S}}}^+(\dot{V})$ we arrive at 2).

2)$\Rightarrow$3): Let $\dot{\Gamma}$ be an open subset of $\partial \dot{V} \cap \overline{\dot{E}}_t^-$ $(t \in I)$ in the induced topology. Since each point of set $\partial \dot{V} \cap \overline{\dot{E}}_t^-$ is regular for $\dot{V}$, the harmonic function $P_{\dot{E} \backslash \dot{V}}(\cdot, 1_{\dot{\Gamma}})$ is strictly positive near $\dot{\Gamma}$. From the Harnack inequality 2) it now follows that it is strictly positive also at point $\dot{x}$.

3)$\Rightarrow$4): Let A be an absorbent subset of harmonic space $(\dot{V}, \dot{\mathcal{S}})$. From 3) it follows that together with each of its points $\dot{x}_0 = (x_0, t_0)$, the set A contains also some semislab $\overline{\dot{W}}_{t_0}^- = W \times \{t \in I : t \leq t_0\}$, where W is some neighborhood of x_0 in $\mathcal{F}$. We set $A_{t_0} := \{x \in V : (x, t_0) \in A\}$. Then A_{t_0} is a nonempty closed subset of V. Consequently, $A_{t_0} = V$, but then the set A also contains semispace $\dot{V} \cap \overline{\dot{E}}_{t_0}^-$. Consequently the set A itself is a semispace.

4)$\Rightarrow$1): For each fixed $y \in V$, the function

$$\dot{x} \to p_V(t; x, y) \quad (\dot{x} = (x, t))$$

is hyperharmonic on the set $\dot{V} = V \times I$, where $I \subset \,]0, +\infty[$. Therefore the set $A = \{\dot{x} \in \dot{V} : p_V(t; x, y) = 0\}$ is absorbent. According to 4) the set A coincides with some semispace, and since $p_V(t; y, y) > 0$ for any $t > 0$, the set A is empty.

In order to conclude the proof of the proposition we note that if assertion 2) is valid for some base $\mathcal{F}_0$, then it is easy to verify that 1) is true for any base $\mathcal{F}$ consisting of connected sets. But then the remaining assertions are true for $\mathcal{F}$. $\Box$

3.3 Projective sequences of harmonic spaces: examples, definitions, statements of theorems

3.3.1

First we shall consider several model examples. We shall systematically use these examples in this section, as well as in Sec. 3.5, for illustration of general definitions and results.

Example 3.3.1. We shall consider a Euclidean space $\mathbb{R}^\alpha$ and the hyperharmonic sheaf $\mathcal{S}_\alpha := \mathcal{S}_{\Delta_\alpha}$ on $\mathbb{R}^\alpha$ generated by Laplace operator $\Delta_\alpha = \sum_{k=1}^{\alpha} \partial_k^2$. As already noted (Secs. 2.5 and 2.6), $(\mathbb{R}^\alpha, \mathcal{S}_\alpha)$ is a harmonic space (and even a $\mathcal{P}$-space, if $\alpha \geq 3$). Let $\pi_{\alpha\beta}$ $(\alpha \leq \beta)$ be the projection of space $\mathbb{R}^\beta$ onto space $\mathbb{R}^\alpha$. We cite a well-known property of classical hyperharmonic functions: for any open set $V \subset \mathbb{R}^\alpha$ and function $u \in \mathcal{S}_\alpha(V)$, the function $\underline{u} := u \cdot \pi_{\alpha\beta}$ defined on set $\underline{V} := \pi_{\alpha\beta}^{-1}(V)$ belongs to $\mathcal{S}_\beta(\underline{V})$ for any $\beta \geq \alpha$. We shall symbolically represent the noted property of hyperharmonic functions in the form

$$\mathcal{S}_\alpha \cdot \pi_{\alpha\beta} \subset \mathcal{S}_\beta \quad (\alpha \leq \beta).$$

Thus we obtain an increasing (in the above-mentioned sense) sequence $\{(\mathbb{R}^\alpha, \mathcal{S}_\alpha), \alpha = 1, 2, \ldots\}$ of harmonic spaces. Is it possible in some sense to close this sequence? ... We postpone discussion of this question until Sec. 3.5 and go to the next examples which we shall examine in this section.

Example 3.3.2. Consider the following objects:
$\mathbb{T}^\alpha$ is the α-dimensional torus,
$\Delta_\alpha = \sum_{k=1}^{\alpha} \partial_k^2$ is the Laplace operator on $\mathbb{T}^\alpha$,
$\mathcal{S}_\alpha := \mathcal{S}_{\Delta_\alpha}$ is the hyperharmonic sheaf on $\mathbb{T}^\alpha$ generated by operator Δ_α,
$\pi_{\alpha\beta} : \mathbb{T}^\beta \to \mathbb{T}^\alpha$ is the projection of $\mathbb{T}^\beta$ onto $\mathbb{T}^\alpha$ $(\alpha \leq \beta)$.

For every $\alpha = 1, 2, \ldots$, $(\mathbb{T}^\alpha, \mathcal{S}_\alpha)$ is a harmonic space and, as in Example 3.3.1, the inclusions

$$\mathcal{S}_\alpha \cdot \pi_{\alpha\beta} \subset \mathcal{S}_\beta \quad (\alpha \leq \beta)$$

are satisfied.

Example 3.3.3. Let $\{a_k\}_1^\infty$ be a sequence of positive numbers. For every $\alpha = 1, 2, \ldots$ we shall consider the objects:

$\mathcal{L}_\alpha = \sum_{k=1}^\alpha a_k \partial_k^2$ is a differential operator on $\mathbb{T}^\alpha$,

$\mathcal{S}_\alpha := \mathcal{S}_{\mathcal{L}_\alpha}$ is the hyperharmonic sheaf on $\mathbb{T}^\alpha$ generated by operator $\mathcal{L}_\alpha$.

Here, as in the preceding examples, for any $\alpha = 1, 2, \ldots$, $(\mathbb{T}^\alpha, \mathcal{S}_\alpha)$ is a harmonic space and the inclusions

$$\mathcal{S}_\alpha \cdot \pi_{\alpha\beta} \subset \mathcal{S}_\beta \quad (\alpha \le \beta)$$

hold.

3.3.2

We pass to the general construction. First we recall the definitions that we shall need in the following. Let $\{E^\alpha\}_1^\infty$ be a sequence of sets and $\{\pi_{\alpha\beta}\}_1^\infty$ the sequence of mappings $\pi_{\alpha\beta} : E^\beta \to E^\alpha$ $(\alpha \le \beta)$. We shall say that $\{E^\alpha; \pi_{\alpha\beta}\}$ is a *projective* sequence of sets (with respect to mappings $\pi_{\alpha\beta}$), if the condition:

$$\pi_{\alpha\gamma} = \pi_{\alpha\beta} \circ \pi_{\beta\gamma} \quad (\forall \alpha \le \beta \le \gamma)$$

is satisfied.

Let $\widetilde{E} = \prod_{\alpha=1}^\infty E^\alpha$ be the direct product of sets E^α and pr_α the projection of set $\widetilde{E}$ onto E^α. We denote by E the subset of $\widetilde{E}$ consisting of elements x satisfying the condition:

$$pr_\alpha(x) = \pi_{\alpha\beta} \circ pr_\beta(x) \quad (\forall \alpha \le \beta).$$

The set E is called the *projective limit* of the projective sequence of sets $\{E^\alpha; \pi_{\alpha\beta}\}$; in this connection we use the notation $E = \lim \text{proj}\, E^\alpha$ or $E = \lim_{\leftarrow} E^\alpha$. The mapping $\pi_\alpha := pr_\alpha|_E$ is called the *canonical* mapping of set E onto E^α. Every subset of E of the type $\pi_\alpha^{-1}(V)$, $V \subset E^\alpha$, is called cylindrical, and the set V is called its basis; the term "cylindrical" is ascribed also to functions of type $u \cdot \pi_\alpha$.

Now for each $\alpha = 1, 2, \ldots$, let $(E^\alpha, \mathfrak{U}_\alpha)$ be a *topological space*. It is said that $\{E^\alpha, \mathfrak{U}_\alpha; \pi_{\alpha\beta}\}_1^\infty$ is a *projective* sequence of topological spaces if the following conditions are satisfied:

1) $\{E^\alpha; \pi_{\alpha\beta}\}_1^\infty$ is a projective sequence of sets;
2) $\forall \alpha \le \beta$, the mapping $\pi_{\alpha\beta}$ is continuous.

The set $E = \lim \text{proj}\, E^\alpha$ provided with the topology $\mathfrak{U}$ which is the weakest of the topologies in which all mappings π_α are continuous, is called the *projective* limit of the projective sequence of topological spaces $\{E^\alpha, \mathfrak{U}_\alpha; \pi_{\alpha\beta}\}_1^\infty$.

Finally, for every $\alpha = 1, 2, \ldots$, $(E^\alpha, \mathcal{S}_\alpha)$ is a *harmonic space*. We shall say that $\{E^\alpha, \mathcal{S}_\alpha; \pi_{\alpha\beta}\}_1^\infty$ is a *projective* sequence of harmonic spaces if the following conditions are satisfied:

1) $\{E^\alpha, \mathcal{U}_\alpha; \pi_{\alpha\beta}\}_1^\infty$ is a projective sequence of topological spaces;
2) $\mathcal{S}_\alpha \cdot \pi_{\alpha\beta} \subset \mathcal{S}_\beta \ (\forall \alpha \le \beta)$,

where the inclusion is understood in the sense that: $\forall V \in \mathcal{U}_\alpha, \forall u \in \mathcal{S}_\alpha(V)$, the function $\underline{u} := u \cdot \pi_{\alpha\beta}$ defined on set $\underline{V} := \pi_{\alpha\beta}^{-1}(V)$ is contained in $\mathcal{S}_\beta(\underline{V})$.

Let $(E, \mathcal{U})$ be the projective limit of a projective sequence of topological spaces $\{E^\alpha, \mathcal{U}_\alpha; \pi_{\alpha\beta}\}_1^\infty$ and $\mathcal{S}$ be *minimal* among all hyperharmonic sheaves $\mathcal{G}$ on E satisfying the conditions:

1) $\forall V \in \mathcal{U}, \forall \{u_n\}_1^\infty \subset \mathcal{G}(V) : u_n \uparrow \Rightarrow \sup_n u_n \in \mathcal{G}(V)$;
2) $\mathcal{S}_\alpha \cdot \pi_\alpha \subset \mathcal{G} \ (\forall \alpha = 1, 2, \dots)$.

We shall call the space $(E, \mathcal{S})$ the *projective* limit of the projective sequence of harmonic spaces $\{E^\alpha, \mathcal{S}_\alpha : \pi_{\alpha\beta}\}_1^\infty$, and in this connection we shall sometimes denote sheaf $\mathcal{S}$ by $\lim \operatorname{proj} \mathcal{S}_\alpha$.

Remark 3.3.1. In the general case the space $(E, \mathcal{S})$ is not harmonic if only because the topological space $(E, \mathcal{U})$ can prove to be not locally compact. This effect takes place in Example 3.3.1: $E = \lim \operatorname{proj} \mathbb{R}^\alpha = \mathbb{R}^\infty$. But even if the topological space $(E, \mathcal{U})$ is locally compact, then it is still possible that $(E, \mathcal{S})$ is not a harmonic space. Below we shall show that this takes place in Example 3.3.2. In contrast to this, the space $(E, \mathcal{S})$ of Example 3.3.3 will be harmonic if the sequence $\{a_k\}$ increases sufficiently rapidly: $\sum_{k=1}^\infty 1/a_k < \infty$. The general case will be analyzed below in Theorems 3.3.1 and 3.3.2 under natural limitations of topological character.

3.3.3

We shall fix a projective sequence of harmonic spaces $\{E^\alpha, \mathcal{S}_\alpha; \pi_{\alpha\beta}\}_1^\infty$ and we shall assume (until Sec. 3.5) that these conditions are satisfied: $\forall \alpha \le \beta$
1) $1 \in \mathcal{S}_\alpha^+(E^\alpha)$;
2) $(E^\alpha, \mathcal{S}_\alpha)$ is a $\mathcal{P}$-space;
3) the mapping $\pi_{\alpha\beta}$ is open, $\pi_{\alpha\beta}(E^\beta) = E^\alpha$;
4) for any compact set $K \subset E^\alpha$, the set $\pi_{\alpha\beta}^{-1}(K)$ is a compact set in E^β.

We shall omit the standard proof of the following proposition.

Proposition 3.3.1. *Let a topological space $(E, \mathcal{U})$ be the projective limit of a projective sequence of topological spaces $\{E^\alpha, \mathcal{U}_\alpha; \pi_{\alpha\beta}\}_1^\infty$ satisfying conditions 3) and 4); then*
1) $(E, \mathcal{U})$ *is a locally compact Hausdorff space with countable base;*
2) *for any function $\varphi \in \mathbb{C}(E)$, for any $\alpha = 1, 2, \dots$ and any $x \in E^\alpha$ we set*

$$\overline{\overline{\varphi}}_\alpha(x) := \inf_{\pi_\alpha^{-1}(x)} \varphi, \quad \varphi_\alpha := \overline{\overline{\varphi}}_\alpha \cdot \pi_\alpha,$$

then these properties hold:

a) $\overline{\overline{\varphi}}_\alpha \in \mathbb{C}(E^\alpha)$;

b) $\varphi_\alpha \uparrow \varphi$, $\alpha \uparrow \infty$.

As already noted (Theorem 2.8.1), with each $\mathcal{P}$-harmonic space in which 1 is hyperharmonic, it is possible to associate a Markov process with a "good" semigroup and strong Feller resolvent[3] and such that the cone of excessive functions generated by them coincides with the cone of non-negative hyperharmonic functions. A similar result takes place also in the situation we are considering. In this connection, in the general case the resolvent will have the strong Feller property only *locally (with respect to* α ...*)*, i.e. it will transform bounded *cylindrical* Borel functions to continuous ones. The *strong Feller* property will be satisfied if and only if $(E, \mathcal{S})$ is a harmonic space.

We go to exact statements. In the theorem which we shall give below, the following notation is used:

$X = (x_t, \xi, \mathcal{M}_t, P^x)$ is a Markov process on $(E, \mathcal{B})$;

$\mathbb{P} = (P_t)_{t>0}$ is the Markov semigroup of process X;

$\mathbb{R} = (R_\lambda)_{\lambda>0}$ is the resolvent of semigroup $\mathbb{P}$

$\mathcal{S}^{\mathbb{P}}$ is the cone of excessive functions with respect to semigroup $\mathbb{P}$;

$\mathcal{S}^{\mathbb{P}}_c \subset \mathcal{S}^{\mathbb{P}}$ is the cone of excessive lower-semicontinuous functions;

$\mathcal{S}^{\mathbb{P}}_\alpha \subset \mathcal{S}^{\mathbb{P}}$ is the cone of excessive π_α-measurable functions;

$\underline{\mathcal{S}}^+_\alpha = \{u = \overline{\overline{u}} \cdot \pi_\alpha : (\exists)\overline{\overline{u}} \in \mathcal{S}^+_\alpha(E^\alpha)\}$;

$\mathbb{S}_{\{\mathcal{S}^+_\alpha\}} = \{u = \uparrow \lim_{\alpha \to \infty} u_\alpha : (\exists)\{u_\alpha\}^\infty_1, \ u_\alpha \in \underline{\mathcal{S}}^+_\alpha, \ u_\alpha \uparrow u\}$;

$\mathbb{B}(E)$ is the set of bounded cylindrical Borel functions.

Theorem 3.3.1. *There exists a standard process X with almost surely continuous trajectories such that:*

1. $\forall \varphi \in \mathbb{C}_k(E)$
   a) $P_t\varphi \in \mathbb{C}_b(E)$ *($\forall t > 0$)*;
   b) $\lim_{t\downarrow 0} \sup_E |P_t\varphi - \varphi| = 0$;
   c) $\forall F \subset E : \overline{F} \cap \operatorname{supp}\varphi = \emptyset$ *we have*

$$\lim_{t\downarrow 0} \frac{1}{t} \sup_F P_t|\varphi| = 0;$$

2. $\forall \varphi \in \mathbb{C}_b(E) \cup \underline{\mathbb{B}}(E)$ *we have*

$$R_\lambda\varphi \in \mathbb{C}_b(E) \ (\forall \lambda \geq 0).$$

3. a) $\mathcal{S}^{\mathbb{P}}_\alpha = \underline{\mathcal{S}}^+_\alpha$ *($\forall \alpha$)*;
   b) $\mathcal{S}^{\mathbb{P}}_c = \mathbb{S}_{\{\mathcal{S}^+_\alpha\}}$.

Remark 3.3.2. For any $\alpha = 1, 2, \ldots$, we denote by $\underline{\mathcal{B}}_\alpha$ the σ-algebra of cylindrical Borel subsets of E with bases in E^α. It is obvious that the current of σ-algebras $\{\underline{\mathcal{B}}_\alpha\}$ increases monotonically and $\bigvee_\alpha \underline{\mathcal{B}}_\alpha = \mathcal{B}$. For each $\alpha = 1, 2, \ldots$ we

---

3 I.e., $R_\lambda : \mathbb{B} \to \mathbb{C}_b$ $(\forall \lambda \geq 0)$

also define a space $\underline{\mathbb{B}}_\alpha(E)$ as the subspace of $\mathbb{B}(E)$ consisting of π_α-measurable functions, i.e. functions of the type $u \cdot \pi_\alpha$, $u \in \mathbb{B}(E^\alpha)$. We have: $\underline{\mathbb{B}}_\alpha(E) \uparrow$ and $\bigcup_\alpha \underline{\mathbb{B}}_\alpha(E) = \underline{\mathbb{B}}(E)$, however, $\overline{\underline{\mathbb{B}}(E)} \neq \mathbb{B}(E)$. Therefore from Property 2 of Theorem 3.3.1, it does not follow that resolvent $\mathbb{R}$ is a strong Feller resolvent.

Example 3.3.4. We shall consider the torus $\mathbb{T}$ as an interval $[-\pi, \pi]$ with identified endpoints; in this connection we shall consider functions on $\mathbb{T}$ as 2π-periodic functions on the line $\mathbb{R}^1$. Let $\overset{\circ}{n}_t$ be a Gaussian measure on the torus with parameters $(0, 2t)$, i.e. the image of Gaussian measures n_t on the line with the same parameters obtained for the natural homomorphism $\gamma : \mathbb{R}^1 \to \mathbb{R}^1/2\pi\mathbb{Z}$.

We have: $\overset{\circ}{n}_t\,(dx) = \overset{\circ}{n}_t\,(x)\,dx$, where

$$\overset{\circ}{n}_t\,(x) = 2\pi \sum_{-\infty}^{\infty} \frac{1}{\sqrt{4\pi t}} \exp\{-(x - 2\pi m)^2/4t\}$$

and dx is the normed Lebesgue measure on the segment $[-\pi, \pi]$.

The family of measures $(\overset{\circ}{n}_t)_{t>0}$ has the semigroup property with respect to convolution, and weak convergence of measures $\overset{\circ}{n}_t \to \varepsilon_0$ takes place for $t \downarrow 0$. For each $t > 0$ we shall consider the linear operator ${}^1P_t f = \overset{\circ}{n}_t * f$, $f \in \mathbb{C}(\mathbb{T})$. From our constructions it follows that the family ${}^1\mathbb{P} = ({}^1P_t)_{t>0}$ is a continuous semigroup of operators in Banach space $\mathbb{C}(\mathbb{T})$, and its infinitesimal operator on smooth functions coincides with differential operator ∂^2.

Let $\mathbb{T}^\alpha$ be an α-dimensional torus; we shall form the Gaussian semigroup of measures $({}^\alpha\mu_t)_{t>0}$ on $(\mathbb{T}^\alpha, \mathscr{B}_\alpha)$ and its corresponding semigroup of operators ${}^\alpha\mathbb{P} = ({}^\alpha P_t)_{t>0}$ on $\mathbb{C}(\mathbb{T}^\alpha)$:

$$ {}^\alpha\mu_t = \overset{\circ}{n}_t \otimes \ldots \otimes \overset{\circ}{n}_t \ \text{ with } \alpha \text{ factors,}$$

$$ {}^\alpha P_t f = {}^\alpha\mu_t * f, \ \ f \in \mathbb{C}(\mathbb{T}^\alpha).$$

As also in the one-dimensional case, the infinitesimal operator of semigroup ${}^\alpha\mathbb{P}$ on smooth functions coincides with differential operator $\Delta_\alpha = \sum_{k=1}^{\alpha} \partial_k^2$.

And finally, let $(\mu_t)_{t>0}$ and $\mathbb{P} = (P_t)_{t>0}$ be the Gaussian semigroup of measures and the semigroup of operators on $(\mathbb{T}^\infty, \mathscr{B})$ and $\mathbb{C}(\mathbb{T}^\infty)$ respectively such that:

$$\mu_t = \overset{\circ}{n}_t \otimes \ldots \otimes \overset{\circ}{n}_t \ldots \text{ countable number of factors,}$$

$$P_t f = \mu_t * f, \ \ f \in \mathbb{C}(\mathbb{T}^\infty).$$

We denote by ${}^\alpha\mathbb{R} = ({}^\alpha R_\lambda)_{\lambda>0}$ and $\mathbb{R} = (R_\lambda)_{\lambda>0}$ the resolvents of semigroups ${}^\alpha\mathbb{P}$ and $\mathbb{P}$ respectively. It is obvious that for each $\alpha = 1, 2, \ldots$ the semigroup ${}^\alpha\mathbb{P}$ and therefore also the resolvent ${}^\alpha\mathbb{R}$ are strong Feller. For any $\alpha = 1, 2, \ldots$ and for any function $f \in \mathbb{B}(\mathbb{T}^\alpha)$, we have:[4]

$$P_t(f \cdot \pi_\alpha) = ({}^\alpha P_t f) \cdot \pi_\alpha, \quad R_\lambda(f \cdot \pi_\alpha) = ({}^\alpha R_\lambda f) \cdot \pi_\alpha,$$

consequently the resolvent $\mathbb{R}$ has property 2 of Theorem 3.3.1. We shall show that resolvent $\mathbb{R}$ does not have the strong Feller property. We set $dx = dx_1 \otimes dx_2 \otimes \dots$, $x = (x_i)_1^\infty \in \mathbb{T}^\infty$, and we note that for any functions $f, g \in \mathbb{B}(\mathbb{T}^\infty)$, the relations

$$\int_{\mathbb{T}^\infty} f(x) P_t g(x)\, dx = \int_{\mathbb{T}^\infty} g(x) P_t f(x)\, dx \quad (\forall t > 0)$$

are satisfied. Finally, analogous relations are also satisfied for operators R_λ, $\lambda > 0$. Let Γ be a Borel subset of $\mathbb{T}^\infty$ having dx-measure zero, then $\forall \varphi \in \mathbb{C}^+(\mathbb{T}^\infty)$ we have

$$\int_{\mathbb{T}^\infty} \varphi(x) R_\lambda 1_\Gamma(x)\, dx = \int_{\mathbb{T}^\infty} 1_\Gamma(x) R_\lambda \varphi(x)\, dx = 0.$$

Consequently the function $R_\lambda 1_\Gamma(x)$ equals zero dx-almost everywhere. Therefore, if the resolvent $\mathbb{R}$ were strong Feller, then for all $\lambda > 0$ and $x \in \mathbb{T}^\infty$ the measure $R_\lambda(x, \cdot) = R_\lambda 1.(x)$ would be absolutely continuous with respect to measure dx. But then by the Fukushima theorem [56] the measures $P_t(x, \cdot) = P_t 1.(x)$ ($x \in \mathbb{T}^\infty$, $t > 0$) would have the same property. In particular, the measure $P_t(0, \cdot) = \mu_t(\cdot) = \overset{o}{n}_t \otimes \overset{o}{n}_t \otimes \dots$ would be absolutely continuous with respect to measure dx. But by Kakutani's theorem [96: Chapter VII, 6, Theorem 3] this measure is singular with respect to measure dx. Consequently the resolvent $\mathbb{R}$ does not have the strong Feller property. $\qquad\square$

Example 3.3.5. Here we carry out constructions analogous to those which were carried out in Example 3.3.2. Thus, the Gaussian semigroup of measures $({}^\alpha\mu_t)_{t>0}$ on $\mathbb{T}^\alpha$ corresponding to differential operator $\mathscr{L}_\alpha = \sum_{k=1}^\alpha a_k \partial_k^2$ has the form

$$^\alpha\mu_t = \bigotimes_{k=1}^\alpha \overset{o}{n}_{a_k t},$$

and the corresponding limit semigroup $(\mu_t)_{t>0}$ on $\mathbb{T}^\infty$ is such that:

$$\mu_t = \bigotimes_{k=1}^\infty \overset{o}{n}_{a_k t}.$$

We are going to show that the semigroup of operators $\mathbb{P} = (P_t)_{t>0}$ generated by semigroup $(\mu_t)_{t>0}$ has the strong Feller property if the series $\sum_{k=1}^\infty 1/a_k$ converges. It is obvious that this implies that the corresponding resolvent $\mathbb{R}$ is strong Feller. We shall show that if the series $\sum_{k=1}^\infty 1/a_k$ converges, then for every $t > 0$ the

---

4   They exist by natural extensions of operators P_t and R_λ to Banach space $\mathbb{B}\left(\mathbb{T}^\infty\right)$.

measure μ_t is absolutely continuous with respect to measure dx. The strong Feller property of the semigroup now follows from the inequality

$$|P_t f(x) - P_t f(y)| \ \leq \ ||f|| \int_{\mathbb{T}^\infty} |\mu_t(z - x) - \mu_t(z - y)| \, dz$$

and the continuity of the shift operator on $L_1(dx)$.

We shall write the Fourier transform of function $\overset{\circ}{n}_t(x)$, $x \in [-\pi, \pi]$

$$\widehat{\overset{\circ}{n}_t}(\theta) = \exp(-t\theta^2), \quad \theta \in \mathbb{Z}.$$

Passing to the inverse Fourier transform, we obtain

$$\overset{\circ}{n}_t(x) \leq \overset{\circ}{n}_t(0) = \sum_{\theta \in \mathbb{Z}} e^{-t\theta^2} = \left(1 + 2\sum_{\theta \geq 1} e^{-t\theta^2}\right) \leq \left(1 + 2\sum_{\theta \geq 1} \frac{1}{t\theta^2}\right) =$$

$$= \left(1 + \frac{\text{const}}{t}\right).$$

The obtained inequality permits us to estimate the function $\,^{\alpha}\mu_t(x)$ - the density of measure $\,^{\alpha}\mu_t$ with respect to measure $\,^{\alpha}dx = dx_1 \otimes \ldots \otimes dx_\alpha$:

$$\,^{\alpha}\mu_t(x) = \prod_{k=1}^{\alpha} \overset{\circ}{n}_{a_k t}(x_k) \leq \prod_{k=1}^{\alpha} \left(1 + \frac{\text{const}}{a_k t}\right) \leq$$

$$\leq \prod_{k=1}^{\infty} \left(1 + \frac{\text{const}}{a_k t}\right) \leq \exp \sum_{k=1}^{\infty} \frac{\text{const}}{a_k t} = \exp \frac{\text{const}}{t} < \infty.$$

Now it remains to note that the function $\,^{\alpha}\mu_t \cdot \pi_\alpha$ is the density of measure $\,^{\alpha}\mu_t = \mu_t|_{\underline{\mathcal{B}}_\alpha}$ with respect to measure $\,^{\alpha}\underline{dx} = dx|_{\underline{\mathcal{B}}_\alpha}$ and to use the general theorem (see [57: Chapter VII, 1], [96: Chapter VII, 6]) on absolute continuity of measures.

We shall consider the Markov process X on E with properties 1–3 of Theorem 3.3.1. Let $\mathcal{S}_c^X$ be the sheaf of lower-semicontinuous locally superharmonic functions of this process. It will be shown (Sec. 3.4.7) that $\mathcal{S}_c^X$ is minimal among all hyperharmonic sheaves $\mathcal{G}$ on E satisfying the conditions:
1) $\forall V \in \mathcal{U}, \forall \{u_n\}_1^\infty \subset \mathcal{G}(V), \, u_n \uparrow \Rightarrow \sup_n u_n \in \mathcal{G}(V)$;
2) $\mathcal{S}_\alpha \cdot \pi_\alpha \subset \mathcal{G} \ (\forall \alpha)$.

Thus, according to Sec. 3.3.2, this sheaf coincides with the sheaf $\mathcal{S} = \lim \text{proj} \, \mathcal{S}_\alpha$.

As already noted, the space $(E, \mathcal{S})$ in the general case does not satisfy all the axioms of a $\mathcal{P}$-harmonic space. Nevertheless it makes sense to list the axioms which are preserved during passage to the projective limit.

Theorem 3.3.2. *The following properties hold:*
1) *(non-degeneracy axiom) The harmonic sheaf $\mathcal{H}_{\mathcal{S}}$ is not degenerate at any point*
   $x \in E$.

2) *(resolutivity axiom) The family of resolutive (with respect to sheaf $\mathcal{S}$) sets forms the base of a topology of E.*

3) *(weakened convergence axiom) The sheaf $\mathcal{H}_{\mathcal{S}}$ is closed with respect to uniform limit passages.*

4) *(completeness axiom) Let V be an open subset of E and u a lower-semicontinuous function on V : $\forall W \subset V$ (W a resolutive set) and $\forall x \in W$ we have $\mu_x^W(u) \leq u(x)$. Then $u \in \mathcal{S}(V)$.*

5) *(axiom of existence of potential) $\forall x \in E$ the potential $p : p(x) > 0$ exists.*

Remark 3.3.3. According to Theorem 3.3.2, the space $(E, \mathcal{S})$ is a $\mathcal{P}$-harmonic space if and only if the sheaf $\mathcal{H}_{\mathcal{S}}$ satisfies the *Bauer convergence property*. Satisfaction of this axiom is equivalent to satisfaction of one of the following two equivalent conditions:

a) $\mathcal{S}_c^{\mathbb{P}} = \mathcal{S}^{\mathbb{P}}$,

b) resolvent $\mathbb{R}$ is strong Feller (see Theorem 3.3.1).

Example 3.3.6. (continuation of Example 3.3.2.) A harmonic space $(\mathbb{T}^{\alpha}, \mathcal{S}_{\alpha})$ is not a $\mathcal{P}$-space for any α, since every function of $\mathcal{S}_{\alpha}^{+}(\mathbb{T}^{\alpha})$ is some constant and consequently every potential is identically zero. Nevertheless for any number $\lambda > 0$, the hyperharmonic sheaf $\mathcal{S}_{\alpha}^{\lambda}$ generated by operator $\Delta_{\alpha} - \lambda I$ will contain nontrivial potentials and therefore $(\mathbb{T}^{\alpha}, \mathcal{S}_{\alpha}^{\lambda})$ will be a $\mathcal{P}$-harmonic space. For definiteness we shall set $\lambda = 1$. It is easy to see that the semigroup $\mathbb{P}^{1}$ and resolvent $\mathbb{R}^{1}$ in this case have the form (see Example 3.3.4):

$$\mathbb{P}^{1} = (e^{-t}P_t)_{t>0}, \quad \mathbb{R}^{1} = (R_{\lambda+1})_{\lambda \geq 0}.$$

Let X^{1} be a Markov process in phase space $(\mathbb{T}^{\infty}, \mathcal{B})$ generated by semigroup $\mathbb{P}^{1}$. This process has all the properties listed in Theorem 3.3.1 and thus generates a limit space $(\mathbb{T}^{\infty}, \mathcal{S}^{1}) : \mathcal{S}^{1} = \lim \text{proj}\, \mathcal{S}_{\alpha}^{1}$. As was shown, resolvent $\mathbb{R}$ and together with it also resolvent $\mathbb{R}^{1}$ do not have the strong Feller property. Consequently, by Remark 3.3.3., the space $(\mathbb{T}^{\infty}, \mathcal{S}^{1})$ is not harmonic. $\qquad\square$

Example 3.3.7. (continuation of Example 3.3.3.) Let the series $\sum_{k=1}^{\infty} 1/a_k$ converge. Then the resolvent $\mathbb{R}^{1}$ corresponding to projective sequence $\{\mathbb{T}^{\alpha}, \mathcal{S}_{\alpha}^{1}; \pi_{\alpha\beta}\}_{1}^{\infty}$, has the strong Feller property (see Example 3.3.5.). Therefore by Remark 3.3.3, the limit space $(\mathbb{T}^{\infty}, \mathcal{S}^{1})$ is harmonic. $\qquad\square$

3.3.4

We shall dwell on one particular case which arises in applications (see [9], [104]). First we shall give the general theorem on transformation of the phase space of a Markov process [54: Theorem 10.13].

Theorem 3.3.3. *Let* $X = (x_t, \xi, \mathcal{M}_t, P^x)$ *be a Markov process in phase space* $(E, \mathcal{B})$ *with transition function* $p(t; x, \Gamma)$*. Let* γ *be a measurable mapping of* $(E, \mathcal{B})$ *into phase space* $(\widetilde{E}, \widetilde{\mathcal{B}})$ *such that* $\gamma(E) = \widetilde{E}$ *and the following conditions are satisfied:*

1) $\forall \Gamma \in \widetilde{\mathcal{B}}$, $\forall x, y \in E : \gamma(x) = \gamma(y)$, $\forall t \geq 0$ *we have*

$$p(t; x, \gamma^{-1}(\Gamma)) := p(t; y, \gamma^{-1}(\Gamma)),$$

2) $\forall t > 0$, $\forall \Gamma \in \widetilde{\mathcal{B}}$, *the function* $\widetilde{p}$ *on* $\widetilde{E}$ *defined by the equality*

$$\widetilde{p}(t; \widetilde{x}, \Gamma) := p(t; x, \gamma^{-1}(\Gamma)) \quad (\forall x \in \gamma^{-1}(\widetilde{x}))$$

is measurable.[5]

We set $\widetilde{x}_t = \gamma(x_t)$ and let $\widetilde{\mathcal{N}}^0 := \sigma\{(\widetilde{x}_t \in \Gamma) : t \geq 0, \Gamma \in \widetilde{\mathcal{B}}\}$. Further, we let $\widetilde{\mathcal{M}}_t = \mathcal{M}_t \cap \widetilde{\mathcal{N}}^0$ and

$$\widetilde{P}^{\widetilde{x}}(A) = P^x(A) \ (x \in \gamma^{-1}(\widetilde{x}), \ A \in \widetilde{\mathcal{N}}^0).$$

The set $\widetilde{X} = (\widetilde{x}_t, \xi, \widetilde{\mathcal{M}}_t, \widetilde{P}^{\widetilde{x}})$ determines a Markov process in phase space $(\widetilde{E}, \widetilde{\mathcal{B}})$ with transition function

$$\widetilde{p}(t; \widetilde{x}, \Gamma) = p\left(t; x, \gamma^{-1}(\Gamma)\right) \ (x \in \gamma^{-1}(\widetilde{x})).$$

If the process X is strictly Markov, then the process $\widetilde{X}$ is also strictly Markov.

Let $(P_t)_{t>0}$ and $(\widetilde{P}_t)_{t>0}$ be semigroups corresponding to processes X and $\widetilde{X}$, and let A and $\widetilde{A}$ be infinitesimal operators of these semigroups. These equalities hold:

$$(\widetilde{P}_t f) \cdot \gamma = P_t(f \cdot \gamma) \quad (\forall f \in \mathbb{B}(\widetilde{E}, \widetilde{\mathcal{B}}))$$
$$(\widetilde{A} f) \cdot \gamma = A(f \cdot \gamma) \quad (\forall f \in \operatorname{dom} \widetilde{A}).$$

It is said that the process $\widetilde{X}$ is obtained from X by means of a transformation of phase space γ, and we write $\widetilde{X} = \gamma(X)$.

Let $\{E^\alpha, \mathcal{U}_\alpha; \pi_{\alpha\beta}\}_1^\infty$ be a projective sequence of topological spaces satisfying conditions 3) and 4) from Section 3.3.2, and $(E, \mathcal{U})$ its projective limit. Also for each α let a Markov process X^α be given in phase space $(E^\alpha, \mathcal{B}_\alpha)$.

We shall call the sequence $\{X^\alpha; \pi_{\alpha\beta}\}_1^\infty$ *projective* if the condition:

$$X^\alpha = \pi_{\alpha\beta}(X^\beta) \ (\forall \alpha, \beta : \alpha \leq \beta)$$

is satisfied.

We shall call a Markov process X in phase space $(E, \mathcal{B})$ the *projective limit* of projective sequence of processes $\{X^\alpha; \pi_{\alpha\beta}\}_1^\infty$ if the condition:

$$X^\alpha = \pi_\alpha(X) \ (\forall \alpha)$$

---

5   In Theorem 10.13 of [54], the condition $\gamma(\mathcal{B}) \subseteq \widetilde{\mathcal{B}}$ is given, from which condition 2) certainly follows, but not conversely. Together with this, the proof of the theorem for condition 2) does not differ much from that given in [54].

is satisfied, where we shall write $X = \lim \operatorname{proj} X^\alpha$ or $X = \underleftarrow{\lim} X^\alpha$.

The existence and uniqueness[6] of this process X is established in the following manner: first the transition function $p(t; x, \Gamma)$ of this process is defined on the algebra of cylindrical sets $\underline{\mathscr{B}} = \cup_\alpha \underline{\mathscr{B}}_\alpha$ by the equation:[7]

$$p(t; x, \Gamma) = p^\alpha(t; \pi_\alpha(x), \pi_\alpha(\Gamma)) \quad (\Gamma \in \underline{\mathscr{B}}_\alpha),$$

and then can be continued to a σ-algebra $\bigvee_\alpha \underline{\mathscr{B}}_\alpha$ which coincides with $\mathscr{B}$; that such a continuation is possible follows, for example, from the local compactness of space E.

Proposition 3.3.2. *Let* $\{X^\alpha; \pi_{\alpha\beta}\}_1^\infty$ *be a projective sequence of Markov processes and X its projective limit. We shall assume that for every α the following conditions are satisfied:*
1) *X^α is a standard process with almost surely continuous trajectories;*
2) *sheaf $\mathscr{S}_\alpha := \mathscr{S}^{X^\alpha}$ of locally superharmonic functions of process X^α is a hyperharmonic sheaf and $(E^\alpha, \mathscr{S}_\alpha)$ is a harmonic space.*

Then $\{E^\alpha, \mathscr{S}_\alpha; \pi_{\alpha\beta}\}_1^\infty$ is a projective sequence of harmonic spaces and, if $\mathscr{S} := \mathscr{S}_c^X$ is the sheaf of lower-semicontinuous locally superharmonic functions of process X, then the space $(E, \mathscr{S})$ is the projective limit of this sequence.

Example 3.3.8. For definiteness we shall consider Example 3.3.2, since Example 3.3.3 is considered completely analogously. In Example 3.3.4 for every α we constructed the semigroup of operators $^\alpha\mathbb{P} = (^\alpha P_t)_{t>0}$ on $\mathbb{C}(\mathbb{T}^\alpha)$, whose infinitesimal operator on smooth functions coincides with differential operator $\Delta_\alpha = \sum_{k=1}^\alpha \partial_k^2$: $^\alpha P_t f = {}^\alpha\mu_t * f$, $f \in \mathbb{C}(\mathbb{T}^\alpha)$, $^\alpha\mu_t = \overset{o}{n}_t \otimes \ldots \otimes \overset{o}{n}_t$ with α factors, $\overset{o}{n}_t$ is the Gaussian measure on $\mathbb{T}$ with parameters $(0, 2t)$. The noted equations permit us to easily verify the truth of the relations:

$$^\beta P_t(f \cdot \pi_{\alpha\beta}) = (^\alpha P_t f) \cdot \pi_{\alpha\beta} \qquad (f \in \mathbb{C}(\mathbb{T}^\alpha), \alpha \le \beta).$$

Now let X^α be a Markov process in phase space $(\mathbb{T}^\alpha, \mathscr{B}_\alpha)$ with semigroup $^\alpha\mathbb{P}$, $\alpha = 1, 2, \ldots$. From the compatibility of semigroups $\{^\alpha\mathbb{P}\}$ it follows that the sequence of processes $\{X^\alpha; \pi_{\alpha\beta}\}_1^\infty$ is projective, and Markov process X with semigroup $\mathbb{P} = (P_t)_{t>0}$:

$$P_t f = \mu_t * f, \ f \in \mathbb{C}(\mathbb{T}^\infty),$$

$$\mu_t = \overset{o}{n}_t \otimes \overset{o}{n}_t \otimes \ldots \text{ countably many factors}$$

is the projective limit of this sequence. Finally we note that

$$\mathscr{S}_{\Delta_\alpha} = \mathscr{S}^{X^\alpha} \ (\forall \alpha),$$

and therefore $\lim \operatorname{proj} \mathscr{S}_\alpha = \mathscr{S}_c^X$ by Proposition 3.3.2. $\qquad\qquad\square$

---

6   We shall not distinguish any two equivalent processes.

7   Relations $X^\alpha = \pi_{\alpha\beta}(X^\beta)$, $\alpha \le \beta$, imply that function $p(t; x, \Gamma)$ is correctly defined.

3.4 Projective sequences of harmonic spaces: proofs of theorems

In this section we shall fix a projective sequence of harmonic spaces $\{E^\alpha, \mathscr{S}_\alpha; \pi_{\alpha\beta}\}_1^\infty$ satisfying the conditions of Sec. 3.3.3, and give the proofs of Theorems 3.3.1 and 3.3.2, and Proposition 3.3.2. We shall give the proof of Theorem 3.3.1 in several stages: Secs. 3.4.1–3.4.10. In the remaining two sections we shall give the proofs of Theorem 3.3.2 and Proposition 3.3.2, respectively.

3.4.1

Let V_α be an open subset of E^α and let $V_\beta := \pi_{\alpha\beta}^{-1}(V_\alpha)$ be an open subset of E^β. Further, let $^\alpha\mu.^{V_\alpha}$ be a harmonic measure of V_α with respect to sheaf $\mathscr{S}_\alpha$, and $^\beta\mu^{V_\beta}$ a harmonic measure of V_β with respect to sheaf $\mathscr{S}_\beta$.

Proposition 3.4.1. *For any function $\varphi \in \mathbb{C}_b(E^\alpha)$ and any $x \in V_\beta$ the equality:*

$$^\beta\mu_x^{V_\beta}(\varphi \cdot \pi_{\alpha\beta}) = \ ^\alpha\mu_{\pi_{\alpha\beta}(x)}^{V_\alpha}(\varphi)$$

holds.

Proof. We set:

$$\overline{\Gamma}_\alpha := \{u \in \mathscr{S}_\alpha(V_\alpha) : \liminf_{x \to y} u(x) \geq \varphi(y), \forall y \in \partial V_\alpha\},$$

$$\underline{\Gamma}_\alpha := \{u \in -\mathscr{S}_\alpha(V_\alpha) : \limsup_{x \to y} u(x) \leq \varphi(y), \forall y \in \partial V_\alpha\},$$

$$\overline{\Gamma}_\alpha \cdot \pi_{\alpha\beta} := \{u \cdot \pi_{\alpha\beta}, u \in \overline{\Gamma}_\alpha\}, \quad \underline{\Gamma}_\alpha \cdot \pi_{\alpha\beta} := \{u \cdot \pi_{\alpha\beta}, u \in \underline{\Gamma}_\alpha\}.$$

We define the sets $\overline{\Gamma}_\beta$ and $\underline{\Gamma}_\beta$ by analogy with $\overline{\Gamma}_\alpha$ and $\underline{\Gamma}_\alpha$, but for the function $\varphi \cdot \pi_{\alpha\beta}$. We have $\overline{\Gamma}_\alpha \cdot \pi_{\alpha\beta} \subset \overline{\Gamma}_\beta$, $\underline{\Gamma}_\alpha \cdot \pi_{\alpha\beta} \subset \underline{\Gamma}_\beta$, from which we get

$$\sup\{g(x) : g \in \underline{\Gamma}_\alpha \cdot \pi_{\alpha\beta}\} \leq \sup\{g(x) : g \in \underline{\Gamma}_\beta\} = \ ^\beta\mu_x^{V_\beta}(\varphi \cdot \pi_{\alpha\beta}) =$$

$$= \inf\{g(x) : g \in \overline{\Gamma}_\beta\} \leq \inf\{g(x) : g \in \overline{\Gamma}_\alpha \cdot \pi_{\alpha\beta}\},$$

$$\sup\{g(x) : g \in \underline{\Gamma}_\alpha \cdot \pi_{\alpha\beta}\} = \ ^\alpha\mu_{\pi_{\alpha\beta}(x)}^{V_\alpha}(\varphi) = \inf\{g(x) : g \in \overline{\Gamma}_\alpha \cdot \pi_{\alpha\beta}\}.$$

$\square$

For every α let a measure $^\alpha\mu$ be given on $(E^\alpha, \mathscr{B}_\alpha)$. We shall say that $\{^\alpha\mu, \pi_{\alpha\beta}\}_1^\infty$ is a projective sequence of measures if the following condition is satisfied:

$$^\beta\mu(\pi_{\alpha\beta}^{-1}(\Gamma)) = \ ^\alpha\mu(\Gamma) \quad (\forall \alpha, \beta : \alpha \leq \beta, \ \forall \Gamma \in \mathscr{B}_\alpha).$$

We shall call the measure μ on $(E, \mathscr{B})$ the projective limit of the sequence $\{^\alpha\mu; \pi_{\alpha\beta}\}_1^\infty$ if the following condition is satisfied:

$$\mu(\pi_\alpha^{-1}(\Gamma)) = \ ^\alpha\mu(\Gamma) \quad (\forall \alpha, \forall \Gamma \in \mathscr{B}_\alpha).$$

By the local compactness of space E, every projective sequence of measures $\{{}^{\alpha}\mu; \pi_{\alpha\beta}\}_1^{\infty}$ has a projective limit μ; we shall use the notation $\mu = \lim \mathrm{proj}^{\alpha} \mu$.

Now let $V = \pi_{\alpha}^{-1}(V_{\alpha})$ be an open cylindrical subset of E, and let $V_{\beta} := \pi_{\beta}(V)$, $\alpha < \beta$, $x \in V$. According to Proposition 3.4.1, the sequence of measures ${}^{\beta}\mu :=$ ${}^{\beta}\mu_{\pi_{\beta}(x)}^{V_{\beta}}$ is projective with respect to the system of mappings $\{\pi_{\alpha\beta}\}$; we set

$$\mu_x^V := \lim \mathrm{proj}^{\beta} \mu_{\pi_{\beta}(x)}^{V_{\beta}}.$$

Thus the family $\{\mu_x^V\}_{V \in \underline{\underline{\mathcal{U}}}, x \in V}$ of measures[8] is defined on $(E, \mathcal{B})$; we shall call each measure of this family a harmonic measure.

3.4.2

Let $\mathcal{P}_{\alpha} \subset \mathcal{S}_{\alpha}^+(E^{\alpha})$ be the cone of potentials on E^{α}; we set

$$\mathcal{P}_{\alpha} \cdot \pi_{\alpha\beta} := \{p \cdot \pi_{\alpha\beta} : p \in \mathcal{P}_{\alpha}\} \subset \mathcal{S}_{\beta}^+(E^{\beta}).$$

Proposition 3.4.2. *For any $\alpha < \beta$, we have the inclusion*

$$\mathcal{P}_{\alpha} \cdot \pi_{\alpha\beta} \subset \mathcal{P}_{\beta}.$$

Proof. Let $p \in \mathcal{P}_{\alpha}$ and $q \in \mathcal{H}_{\mathcal{S}_{\beta}}^+(E^{\beta})$ such that $0 \leq q \leq p \cdot \pi_{\alpha\beta}$. We set

$$\widetilde{q}(x) := \sup_{\pi_{\alpha\beta}^{-1}(x)} q(y) \quad (x \in E^{\alpha}),$$

then $\widetilde{q} \in \mathbb{C}^+(E^{\alpha})$ and in addition,

$$0 \leq q \leq \widetilde{q} \cdot \pi_{\alpha\beta} \leq p \cdot \pi_{\alpha\beta}.$$

We shall show that $\widetilde{q} \in (-\mathcal{S}_{\alpha})(E^{\alpha})$. From this fact and from the inequality $0 \leq \widetilde{q} \leq p$ it will follow that $\widetilde{q} \equiv 0$, but then also $q \equiv 0$; consequently, $p \cdot \pi_{\alpha\beta} \in \mathcal{P}_{\beta}$. Let V be an open subset of E^{α}, $x \in V$, and $y \in \pi_{\alpha\beta}^{-1}(x)$; then by Proposition 3.4.1 we shall have

$$^{\alpha}\mu_x^V(\widetilde{q}) = {}^{\beta}\mu_y^{\pi_{\alpha\beta}^{-1}(V)}(\widetilde{q} \cdot \pi_{\alpha\beta}) \geq {}^{\beta}\mu_y^{\pi_{\alpha\beta}^{-1}(V)}(q) = q(y),$$

consequently

$$^{\alpha}\mu_x^V(\widetilde{q}) \geq \sup_{\pi_{\alpha\beta}^{-1}(x)} q(y) = \widetilde{q}(x).$$

In view of the arbitrariness of $V \subset E^{\alpha}$, $x \in V$, and the continuity of function $\widetilde{q}$, by the axiom of completeness, we conclude that $\widetilde{q} \in (-\mathcal{S}_{\alpha})(E^{\alpha})$. $\square$

---

8 $\underline{\underline{\mathcal{U}}}$ is the family of open cylindrical subsets of E.

3.4.3

For $\alpha = 1, 2, 3, \ldots$ we set

$$
\begin{aligned}
P_\alpha &:= \mathscr{P}_\alpha \cap \mathbb{C}(E^\alpha), \\
\underline{P}_\alpha &:= \{p \cdot \pi_\alpha : p \in P_\alpha\}, \\
\underline{\mathbb{B}}_\alpha &:= \{u \cdot \pi_\alpha : u \in \mathbb{B}(E^\alpha)\}.
\end{aligned}
$$

Proposition 3.4.3. *For any potential $p \in \underline{P}_\alpha$ there exists a unique resolvent $\mathbb{R} = (R_\lambda)_{\lambda > 0}$ on $\mathbb{B}$ such that*
1) $R_0 1 = p$,
2) $R_0(\underline{\mathbb{B}}_\beta^+) \subset \underline{P}_\beta$ $(\forall \beta \geq \alpha)$,
3) $R_\lambda(\mathbb{C}_b(E)) \subset \mathbb{C}_b(E)$ $(\forall \lambda \geq 0)$.

Proof. Let $p = p_\alpha \cdot \pi_\alpha$, $p_\alpha \in P_\alpha$. According to [46: Proposition 10.2.2] there exists a unique resolvent $\mathbb{R}^{(\alpha)} = (R_\lambda^{(\alpha)})_{\lambda \geq 0}$ on $\mathbb{B}(E^\alpha)$ such that:
1) $R_0^{(\alpha)} 1 = p_\alpha$,
2) $R_0^{(\alpha)}(\mathbb{B}^+(E^\alpha)) \subset P_\alpha$,
3) $R_\lambda^{(\alpha)}(\mathbb{C}_b(E^\alpha)) \subset \mathbb{C}_b(E^\alpha)$ $(\forall \lambda > 0)$.

We set $p_\beta = p_\alpha \cdot \pi_{\alpha\beta}$, $\beta > \alpha$. Then $p = p_\beta \cdot \pi_\beta$ and by Proposition 3.4.2 we have $p_\beta \in P_\beta$. Now let $\mathbb{R}^{(\beta)}$ be the resolvent on $\mathbb{B}(E^\beta)$ corresponding to potential p_β. We shall show that the resolvents $\mathbb{R}^{(\alpha)}$ and $\mathbb{R}^{(\beta)}$ agree with mapping $\pi_{\alpha\beta}$, i.e.

$$
R_\lambda^{(\beta)}(\varphi \cdot \pi_{\alpha\beta}) = (R_\lambda^{(\alpha)}\varphi) \cdot \pi_{\alpha\beta} \quad (\forall \lambda \geq 0, \ \varphi \in \mathbb{C}(E^\alpha)).
$$

It is sufficient to verify this relation for $\lambda = 0$. Here we must recall the general construction of the potential kernel R_0. Let $(\mathscr{E}, \mathscr{G})$ be a $\mathscr{P}$-harmonic space and $p \in \mathscr{P}$ a finite potential at all points of $\mathscr{E}$. The finite set $\delta = \{q_1, \ldots, q_n : q_i \in \mathscr{P}\}$ is called a partition of potential p if $p = q_1 + \ldots + q_n$; the set of all partitions is denoted by Δ_p. Now let $\varphi \in \mathbb{C}_k(\mathscr{E})$, and for a given $\delta \in \Delta_p$ we set

$$
\overline{R}^\delta \varphi := \sum_{i=1}^{n} q_i \sup_{S(q_i)} \varphi,
$$

$$
\underline{R}^\delta \varphi := \sum_{i=1}^{n} q_i \inf_{S(q_i)} \varphi,
$$

where $S(q.)$ is the harmonic support of potential $q..$ It is proved [2: Theorem 3.2], [46: Proposition 8.1.5] that

$$
\sup_{\Delta_p} \underline{R}^\delta \varphi = \inf_{\Delta_p} \overline{R}^\delta \varphi;
$$

the common value of the right- and left-hand sides of this equation is by definition the value on the function φ of potential operator R_0 associated with potential p. It is proved as well that the operator R_0 has the properties:

1) $R_0 1 = p$,
2) $R_0(\mathbb{B}^+(\mathcal{E})) \subset \mathcal{P}$,
and is determined uniquely by them.

Now we return to the situation of interest to us and prove that

$$R_0^{(\beta)}(\varphi \cdot \pi_{\alpha\beta}) = (R_0^{(\alpha)}\varphi) \cdot \pi_{\alpha\beta} \quad (\forall \varphi \in \mathbb{C}_k(E^\alpha)).$$

It is obviously sufficient to consider the case $\varphi \geq 0$. Let $\delta_\alpha = \{q_1^\alpha, \ldots, q_n^\alpha : q_i^\alpha \in P_\alpha\}$ be a partition of potential p_α, and let $q_i^\beta = q_i^\alpha \cdot \pi_{\alpha\beta}$, $i = 1, \ldots, n$, and $\delta_\beta = \{q_1^\beta, \ldots, q_n^\beta\}$. It is obvious that δ_β is a partition of potential p_β. We note further that $\pi_{\alpha\beta}(S(q_i^\beta)) = S(q_i^\alpha)$, $i = 1, \ldots, n$, and therefore

$$\begin{aligned}
(\overline{R}^{\delta_\alpha}\varphi) \cdot \pi_{\alpha\beta} &= \sum_{i=1}^{n} q_i^\alpha \cdot \pi_{\alpha\beta} \sup_{S(q_i^\alpha)} \varphi = \\
&= \sum_{i=1}^{n} q_i^\beta \sup_{\pi_{\alpha\beta}(S(q_i^\beta))} \varphi = \\
&= \sum_{i=1}^{n} q_i^\beta \sup_{S(q_i^\beta)} \varphi \cdot \pi_{\alpha\beta} = \overline{R}^{\delta_\beta}(\varphi \cdot \pi_{\alpha\beta}).
\end{aligned}$$

Analogously

$$(\underline{R}^{\delta_\alpha}\varphi) \cdot \pi_{\alpha\beta} = \underline{R}^{\delta_\beta}(\varphi \cdot \pi_{\alpha\beta}).$$

Now we write the chain of inequalities:

$$\begin{aligned}
(^\alpha R_0\varphi) \cdot \pi_{\alpha\beta} &= \sup_{\Delta_{p_\alpha}}(\underline{R}^{\delta_\alpha}\varphi) \cdot \pi_{\alpha\beta} = \sup_{\Delta_{p_\alpha}} \underline{R}^{\delta_\beta}(\varphi \cdot \pi_{\alpha\beta}) \leq \\
&\leq \sup_{\Delta_{p_\beta}} \underline{R}^{\delta}(\varphi \cdot \pi_{\alpha\beta}) = {}^\beta R_0(\varphi \cdot \pi_{\alpha\beta}) = \\
&= \inf_{\Delta_{p_\beta}} \overline{R}^{\delta}(\varphi \cdot \pi_{\alpha\beta}) \leq \inf_{\Delta_{p_\alpha}} \overline{R}^{\delta_\beta}(\varphi \cdot \pi_{\alpha\beta}) = \\
&= \inf_{\Delta_{p_\alpha}}(\overline{R}^{\delta_\alpha}\varphi) \cdot \pi_{\alpha\beta} = (R_0^{(\alpha)}\varphi) \cdot \pi_{\alpha\beta}.
\end{aligned}$$

Thus the operators $R_0^{(\alpha)}$ and $R_0^{(\beta)}$, and also operators $R_\lambda^{(\alpha)}$ and $R_\lambda^{(\beta)}$, $\lambda > 0$, agree with respect to mapping $\pi_{\alpha\beta}$. From this it follows that for every $\lambda \geq 0$ and $x \in E$, the sequence of measures $^\alpha\mu := R_\lambda^{(\alpha)}(x^\alpha, \cdot)$, where $x^\alpha = \pi_\alpha(x)$ and $R_\lambda^{(\alpha)}(x^\alpha, \Gamma) = R_\lambda^{(\alpha)} 1_\Gamma(x^\alpha)$, $\Gamma \in \mathcal{B}_\alpha$, is projective with respect to system of mappings $\{\pi_{\alpha\beta}\}$. We set

$$R_\lambda(x, \cdot) := \lim\mathrm{proj}\, ^\alpha\mu, \quad R_\lambda\varphi(x) := R_\lambda(x, \varphi) \quad (x \in E, \varphi \in \mathbb{B}).$$

It is easy to verify that $\mathbb{R} = (R_\lambda)_{\lambda \geq 0}$ is the resolvent on $\mathbb{B}$ satisfying properties 1) and 2). We shall show that property 3) is satisfied. Let $\underline{\mathbb{C}}_b(E)$ be the subset of $\mathbb{C}_b(E)$ consisting of cylindrical functions; then from compatibility it follows that

$R_\lambda : \underline{\underline{\mathbb{C}}}_b(E) \to \mathbb{C}_b(E)$. Now let $\varphi \in \mathbb{C}_b(E)$, and for every α we set

$$\overline{\overline{\varphi}}_\alpha(x) := \inf_{\pi_\alpha^{-1}(x)} \varphi(y), \quad x \in E^\alpha,$$

$$\varphi_\alpha(x) := \overline{\overline{\varphi}}_\alpha \cdot \pi_\alpha(x), \quad x \in E,$$

then by Proposition 3.3.1 we have $\{\varphi_\alpha\} \subset \underline{\underline{\mathbb{C}}}_b(E)$ and $\varphi_\alpha \uparrow \varphi$. Consequently $R_\lambda\varphi_\alpha \uparrow R_\lambda\varphi$ and thus the function $R_\lambda\varphi$ is lower-semicontinuous. Replacing φ by $-\varphi$ we get that function $R_\lambda(-\varphi)$ is also lower-semicontinuous. Therefore $R_\lambda\varphi$ is a continuous function. $\qquad\square$

3.4.4

We turn first to the general theory. Let $(\mathcal{E}, \mathcal{G})$ be a $\mathcal{P}$-harmonic space. A potential $p \in \mathcal{P}$ is called *hyperstrict* [46: p. 167] if: $\forall \varphi \in \mathbb{C}_k^+(\mathcal{E})$, $\forall \varepsilon > 0$, $\exists p', p'' \in P = \mathcal{P} \cap \mathbb{C}(\mathcal{E})$:

1) [9] $(\exists \delta > 0)$ $p' + p'' \leq \delta p$,
2) $S(p') \cup S(p'') \subset \operatorname{supp} \varphi$;
3) $0 \leq p'' - p' \leq \varphi \leq p'' - p' + \varepsilon$.

According to [46: Proposition 7.2.1] for any strictly positive hyperharmonic function u on $\mathcal{E}$ there exists a hyperstrict potential $p \in P$ such that $p \leq u$.

Now we shall return to the situation which we have been considering. We shall choose a sequence $\{p_\alpha\}_1^\infty \subset \underline{P}$ such that $\forall \alpha$:

1) $p_\alpha = \overline{\overline{p}}_\alpha \cdot \pi_\alpha, \ \overline{\overline{p}}_\alpha \in P_\alpha$;
2) $\overline{\overline{p}}_\alpha$ is a hyperstrict potential (with respect to $(E^\alpha, \mathcal{G}_\alpha)$);
3) $p_{\alpha+1} \leq 1/2 \, p_\alpha, \ p_1 \leq 1/2$,

and we set $p = \sum_{\alpha=1}^\infty p_\alpha$.

Proposition 3.4.4. *There exists a resolvent* $\mathbb{R} = (R_\lambda)_{\lambda \geq 0}$ *on* $\mathbb{B}(E)$ *such that*
1) $R_0 1 = p$,
2) $R_\lambda : \mathbb{C}_b(E) \cup \underline{\underline{\mathbb{B}}}(E) \to \mathbb{C}_b(E)$ *(* $\forall \lambda \geq 0$ *)*.

Proof. For every $\alpha = 1, 2, \ldots$, let R_0^α be the kernel associated in the sense of Proposition 3.4.3 with potential p_α. We set

$$R_0 := \sum_{\alpha=1}^\infty R_0^\alpha.$$

Then it is not difficult to see that the following properties hold:

---

9    $\preceq$ is the specific order on cone $\mathcal{G}^+(\mathcal{E})$: $(\forall f, g \in \mathcal{G}^+(\mathcal{E}))\ f \preceq g \Leftrightarrow \exists d \in \mathcal{G}^+(\mathcal{E}) : g = f + d$ [46: Chapter 5].

1) $R_0 1 = p$,

2) $R_0 : \mathbb{C}_b(E) \cup \underline{\mathbb{B}}(E) \to \mathbb{C}_b(E)$.

We shall show that the kernel R_0 satisfies the *complete maximum principle*, i.e. $\forall f, g \in \mathbb{B}^+(E)$ the relation

$$R_0 g(x) \leq R_0 f(x) + 1 \ (\forall x : g(x) > 0)$$

implies in turn the relation

$$R_0 g(x) \leq R_0 f(x) + 1 \ (\forall x \in E).$$

According to [81: Theorem 4, p. 249] it is sufficient to assume that $f, g \in \mathbb{C}_k^+(E)$. For fixed $\varepsilon > 0$ we set $g_\varepsilon = g/(1+\varepsilon)$, $f_\varepsilon = f/(1+\varepsilon)$ and we write the inequality

$$R_0 g_\varepsilon(x) < R_0 f_\varepsilon(x) + 1, \ \forall x \in \operatorname{supp} g_\varepsilon \ (= \operatorname{supp} g).$$

For every β we consider the sectioned kernel ${}^\beta R_0$:

$$ {}^\beta R_0 := \sum_{\alpha \leq \beta} R_0^\alpha .$$

For the given $\varepsilon > 0$ we choose a number N_ε such that $\forall \beta \geq N_\varepsilon$ the following inequality is satisfied:

$$ {}^\beta R_0 g_\varepsilon(x) \leq \ {}^\beta R_0 f_\varepsilon(x) + 1, \ \forall x \in \operatorname{supp} g_\varepsilon.$$

Now we note that the kernel ${}^\beta R_0$ in particular coincides for $\lambda = 0$ with the value of resolvent ${}^\beta \mathbb{R} = ({}^\beta R_\lambda)$ for potential ${}^\beta p = \sum_{\alpha \leq \beta} p_\alpha$ (see Proposition 3.4.3). Therefore, according to [81: Theorem 68, p. 243], the kernel ${}^\beta R_0$ satisfies the complete maximum principle. Consequently

$$ {}^\beta R_0 g_\varepsilon(x) \leq \ {}^\beta R_0 f_\varepsilon(x) + 1, \ \forall x \in E.$$

Passing to the limit in the obtained inequality for $\beta \uparrow \infty$ and $\varepsilon \downarrow 0$, we obtain

$$R_0 g(x) \leq R_0 f(x) + 1, \ \forall x \in E.$$

Thus is it proved that the kernel R_0 satisfies the complete maximum principle. According to [81: Theorem 1, p. 255] there exists a resolvent $\mathbb{R}$ on $\mathbb{B}(E)$ whose value for $\lambda = 0$ coincides with the kernel $R_0 : \mathbb{R} = (R_\lambda)_{\lambda \geq 0}$. Analysis of the proof of the mentioned theorem from [81] shows that condition 2), satisfied for $\lambda = 0$, is satisfied also for all $\lambda \geq 0$. $\qquad \square$

3.4.5

In this and the following points of this section we shall fix a sequence of hyperstrict potentials $\{p_\alpha\}_1^\infty \subset \underline{P}$, their sum $p = \sum_{\alpha=1}^\infty p_\alpha$ and corresponding resolvent $\mathbb{R} = (R_\lambda)_{\lambda \geq 0}$. We shall use the following notation:

$\mathcal{S}^{\mathbb{R}}$ is the cone of excessive functions with respect to resolvent $\mathbb{R}$, i.e. the set of universally measurable non-negative functions u on E such that $(\forall \lambda > 0)\ \lambda R_\lambda u \leq u$ and $\lambda R_\lambda u \uparrow u$, $\lambda \uparrow \infty$;

$$\underline{\mathcal{S}}_\alpha^+ = \{u = \overline{\overline{u}} \cdot \pi_\alpha : (\exists)\ \overline{\overline{u}} \in \mathcal{S}_\alpha^+(E^\alpha)\},$$

$$\mathbb{S}_{\{\mathcal{S}_\alpha^+\}} = \{u = \uparrow \lim_{\alpha \to \infty} u_\alpha : (\exists)\ \{u_\alpha\}_1^\infty,\ u_\alpha \in \underline{\mathcal{S}}_\alpha^+,\ u_\alpha \uparrow\}.$$

Proposition 3.4.5. $\mathbb{S}_{\{\mathcal{S}_\alpha^+\}} \subset \mathcal{S}^{\mathbb{R}}$.

Proof. (a) Let $u \in \mathbb{S}_{\{\mathcal{S}_\alpha^+\}}$; we shall show that[10]

$$\lambda R_\lambda u \leq u \quad (\forall \lambda > 0).$$

Obviously it is sufficient to consider the case: $u \in \underline{P}_\alpha \cap \mathbb{C}_b(E)$ for some α. According to [81: Theorem 70, p. 245], supermediancy of function u is equivalent to satisfaction of the following condition: for any measurable function h with finite potential $R_0 h$ the relation

$$u(x) \geq R_0 h(x) \quad (\forall x : h(x) > 0)$$

implies the inequality

$$u(x) \geq R_0 h(x) \quad (\forall x \in E).$$

We shall prove that this condition is satisfied. It is sufficient to consider the case: $h \in \mathbb{B}(E)$. We have

$$u(x) + R_0 h^-(x) \geq R_0 h^+(x) \quad (\forall x : h^+(x) > 0).$$

For a number $\varepsilon > 0$ we set $u_\varepsilon = u + \varepsilon$, and then all the more

$$u_\varepsilon(x) + R_0 h^-(x) > R_0 h^+(x) \quad (\forall x : h^+(x) > 0).$$

We choose $N_\varepsilon > \alpha : \forall \beta > N_\varepsilon$

$$u_\varepsilon(x) + {}^\beta R_0 h^-(x) \geq {}^\beta R_0 h^+(x) \quad (\forall x : h^+(x) > 0),$$

where ${}^\beta R_0$ is the truncated kernel corresponding to potential ${}^\beta p = \sum_{\gamma \leq \beta} p_\gamma$. Now we note that $u_\varepsilon \in \underline{\mathcal{S}}_\alpha^+ \subset \underline{\mathcal{S}}_\beta^+$ and ${}^\beta p \in \underline{P}_\beta$, consequently $u_\varepsilon = \overline{\overline{u}}_\varepsilon \cdot \pi_\beta$ and ${}^\beta p = {}^\beta \overline{p} \cdot \pi_\beta$, where $\overline{\overline{u}}_\varepsilon \in \mathcal{S}_\beta^+(E^\beta)$ and ${}^\beta \overline{p} \in P_\beta$, where the potential ${}^\beta \overline{\overline{p}}$ is hyperstrict. According to [46: Proposition 10.2.2] a resolvent $\mathbb{V} = (V_\lambda)_{\lambda \geq 0}$ corresponds to this potential on $\mathbb{B}(E^\beta)$ such that
1) $V_0 1 = {}^\beta \overline{\overline{p}}$,
2) $\mathcal{S}_\beta^+(E^\beta) = \mathcal{S}^{\mathbb{V}}$;
in particular the function $\underline{u}_\varepsilon$ is excessive with respect to this resolvent. On the other hand, by Proposition 3.4.3, a resolvent ${}^\beta \mathbb{R} = ({}^\beta R_\lambda)_{\lambda \geq 0}$ on $\mathbb{B}(E)$ corresponds to a potential ${}^\beta p$ such that

$$ {}^\beta R_\lambda(\varphi \cdot \pi_\beta) = (V_\lambda \varphi) \cdot \pi_\beta \quad (\forall \lambda \geq 0,\ \varphi \in \mathbb{B}(E^\beta)).$$

---

10 Such functions are called *supermedian* (relative to the given resolvent).

Consequently the function $u_\varepsilon = \bar{\bar{u}}_\varepsilon \cdot \pi_\beta$ is excessive with respect to resolvent ${}^\beta\mathbb{R}$, and thus it is also a supermedian. By the above-given criterion of supermediancy we have

$$u_\varepsilon(x) + {}^\beta R_0 h^-(x) \geq {}^\beta R_0 h^+(x) \quad (\forall x \in E).$$

Passing to the limit in this inequality, for $\beta \uparrow \infty$ and $\varepsilon \downarrow 0$ we get

$$u(x) + R_0 h^-(x) \geq R_0 h^+(x) \quad (\forall x \in E).$$

Thus it is proved that u is a supermedian function.

(b) We shall show that

$$\lambda R_\lambda u \uparrow u, \quad \lambda \uparrow \infty \quad (\forall u \in \mathbb{S}_{\{\mathscr{S}_\alpha^+\}}).$$

It is obviously sufficient to establish this property for the functions $u \in \underline{\underline{P}}$.

1. We fix $x \in E$ and set

$$\varphi(u) = \lim_{\lambda \to \infty} \lambda R_\lambda u(x) \quad (\leq u(x)) \quad (u \in \underline{\underline{P}}).$$

We shall show that there exists a measure μ on E such that

$$\varphi(u) = \int u \, d\mu \quad (\forall u \in \underline{\underline{P}}).$$

We fix α and consider the functional φ on the set $\underline{\underline{P}}_\alpha = \{\bar{\bar{u}} \cdot \pi_\alpha, \bar{\bar{u}} \in P_\alpha\}$. It is obvious that φ can be considered as a functional $\bar{\bar{\varphi}}$ on convex cone P_α:

$$\bar{\bar{\varphi}}(\bar{\bar{u}}) := \varphi(u) \quad (u = \bar{\bar{u}} \cdot \pi_\alpha, \ \bar{\bar{u}} \in P_\alpha).$$

This functional is *additive and increases* when $\bar{\bar{u}}$ does, and in addition

$$\bar{\bar{\varphi}}(\bar{\bar{u}}) \leq \bar{\bar{u}}(x_\alpha) \quad (x_\alpha = \pi_\alpha(x)).$$

From this inequality and [46: Proposition 7.1.1] it follows that this functional satisfies one more condition:

$$\inf_{\mathscr{F}_{\bar{\bar{u}}}} \bar{\bar{\varphi}}(\bar{\bar{u}}') = 0,$$

where $\mathscr{F}_{\bar{\bar{u}}}$ is the set of all functions $\bar{\bar{u}}' \in P_\alpha$ for which the sets $\{y \in E^\alpha : \bar{\bar{u}}'(y) < \bar{\bar{u}}(y)\}$ are relatively compact. Therefore according to [46: Lemma, p. 160] there exists a unique measure μ_α on E^α such that

$$\bar{\bar{\varphi}}(\bar{\bar{u}}) = \int \bar{\bar{u}} \, d\mu_\alpha \quad (\forall \bar{\bar{u}} \in P_\alpha).$$

Thus

$$\varphi(u) = \int \bar{\bar{u}} \, d\mu_\alpha \quad (\forall u \in \underline{\underline{P}}_\alpha, \ u = \bar{\bar{u}} \cdot \pi_\alpha).$$

We note further that $\underline{P}_\alpha \subset \underline{P}_\beta$ for $\alpha < \beta$, and therefore

$$\varphi(u) = \int \bar{\bar{u}} \cdot \pi_{\alpha\beta} \, d\mu_\beta \quad (\forall \bar{\bar{u}} \in P_\alpha).$$

Thus we have

$$\int \bar{\bar{u}} d\mu_\alpha = \int \bar{\bar{u}} \cdot \pi_{\alpha\beta} \, d\mu_\beta \quad (\forall \bar{\bar{u}} \in P_\alpha).$$

According to the Deny–Hervé theorem [46: Theorem 2.3.1] the set $P_\alpha - P_\alpha$ is dense in $\mathbb{C}_k(E^\alpha)$, and therefore the above-noted equality holds for any function $\bar{\bar{u}} \in \mathbb{C}_k(E^\alpha)$. Consequently the sequence of measures $\{\mu_\alpha\}_1^\infty$ is projective with respect to the system of mappings $\{\pi_{\alpha\beta}\}_1^\infty$; let μ be its projective limit. It is obvious that

$$\varphi(u) = \int u \, d\mu \quad (\forall u \in \underline{P} = \cup_\alpha \underline{P}_\alpha).$$

2. We shall show that $\forall u \in \mathbb{S}_{\{\mathscr{S}_\alpha^+\}}$

$$\int u \, d\mu = \lim_{\lambda \to \infty} \lambda R_\lambda u(x) \leq u(x).$$

Let $\{u_n\}_1^\infty \subset \underline{P} : u_n \uparrow u$, then

$$u(x) \geq \sup_\lambda \lambda R_\lambda u(x) = \sup_\lambda \sup_n \lambda R_\lambda u_n(x) =$$

$$= \sup_n \sup_\lambda \lambda R_\lambda u_n(x) = \sup_n \int u_n \, d\mu = \int u \, d\mu.$$

3. We shall show that $p(x) = \int p \, d\mu$. In fact, since $p = \sum_{\alpha=1}^\infty p_\alpha$, then $p \in \mathbb{S}_{\{\mathscr{S}_\alpha^+\}}$. Consequently, according to step 2 we will have

$$\int p \, d\mu = \lim_{\lambda \to \infty} \lambda R_\lambda p(x).$$

On the other hand, $p = R_0 1$ is an excessive function, therefore the right-hand side of the equality equals $p(x)$.

4. Set $\nu := \varepsilon_x$. According to points 2 and 3 we have

$$\int p \, d\mu = \int p \, d\nu,$$

$$\int u \, d\mu \leq \int u \, d\nu \quad (\forall u \in \mathbb{S}_{\{\mathscr{S}_\alpha^+\}}).$$

We shall show that $\mu = \nu$. Thus the proof of (b) will be finished, and together with it also the proof of Proposition 3.4.5.

Let $\varphi \in \mathbb{C}_k^+(E^\alpha)$ and $\varepsilon > 0$. We have $p_\alpha = \bar{\bar{p}}_\alpha \cdot \pi_\alpha$, where $\bar{\bar{p}}_\alpha \in P_\alpha$ is a hyperstrict potential. We choose $\bar{\bar{p}}', \bar{\bar{p}}'' \in P_\alpha$ such that:

(1) $(\exists c)$ $\overline{\overline{p}}' + \overline{\overline{p}}'' \preceq c\overline{\overline{p}}_\alpha$,
(2) $0 \le \overline{\overline{p}}'' - \overline{\overline{p}}' \le \varphi \le \overline{\overline{p}}'' - \overline{\overline{p}}' + \varepsilon$,

then $p = \sum_{\alpha=1}^{\infty} p_\alpha = p' + p'' + u$, where the functions $p' = \overline{\overline{p}}' \cdot \pi_\alpha$, $p'' = \overline{\overline{p}}'' \cdot \pi_\alpha$ and u are obviously in $\mathbb{S}_{\{\mathscr{S}_\alpha^+\}}$. We write the inequalities

$$\int p\, d\mu = \int p'\, d\mu + \int p''\, d\mu + \int u\, d\mu \le$$
$$\le \int p'\, d\nu + \int p''\, d\nu + \int u\, d\nu = \int p\, d\nu.$$

From this we obtain

$$\int p'\, d\mu = \int p'\, d\nu, \quad \int p''\, d\mu = \int p''\, d\nu$$

or

$$A := \int (p'' - p')\, d\mu = \int (p'' - p')\, d\nu.$$

From (2) we now obtain

$$A \le \int \varphi \cdot \pi_\alpha\, d\mu \le A + \varepsilon\mu(1),$$
$$A \le \int \varphi \cdot \pi_\alpha\, d\nu \le A + \varepsilon\nu(1),$$

and therefore

$$\left| \int \varphi \cdot \pi_\alpha\, d\mu - \int \varphi \cdot \pi_\alpha\, d\nu \right| \le \varepsilon \max\{\mu(1), \nu(1)\}.$$

By the arbitrariness of $\varepsilon > 0$, we conclude that

$$\int \varphi \cdot \pi_\alpha\, d\mu = \int \varphi \cdot \pi_\alpha\, d\nu$$

By the arbitrariness of $\varphi \in \mathbb{C}_k^+(E^\alpha)$ we conclude that

$$\mu|_{\underline{\underline{\mathscr{B}}}_\alpha} = \nu|_{\underline{\underline{\mathscr{B}}}_\alpha},$$

where $\underline{\underline{\mathscr{B}}}_\alpha$ is the σ-algebra of cylindrical subsets of E with bases in $\mathscr{B}(E^\alpha)$. And finally, by the arbitrariness of α and the fact that $\mathscr{B} = \bigvee_\alpha \underline{\underline{\mathscr{B}}}_\alpha$, we conclude that $\mu = \nu$. $\qquad\qquad\square$

3.4.6

Here we consider the same objects as in the previous section: a sequence of hyperstrict potentials $\{p_\alpha\}_1^\infty$, $p = \sum_{\alpha=1}^{\infty} p_\alpha$ and a resolvent $\mathbb{R} = (R_\lambda)_{\lambda \ge 0}$ on $\mathbb{B}(E)$ such that $R_0 1 = p$.

Proposition 3.4.6. *There exists a standard process* $X = (x_t, \xi, \mathcal{M}_t, P^x)$ *on the phase space* $(E, \mathcal{B})$ *such that*

$$R_\lambda \varphi(x) = M^x \left(\int_0^\xi e^{-\lambda t} \varphi(x_t)\, dt \right) \quad (\forall \lambda \geq 0,\ \forall \varphi \in \mathbb{B}(E)).$$

Proof. 1) We shall show that the property

$$\lim_{\lambda \longrightarrow \infty} \lambda R_\lambda \varphi(x) = \varphi(x) \quad (\forall \varphi \in \mathbb{C}_k(E),\ \forall x \in E)$$

holds. Let $\mathcal{A}$ be the set of functions on E for which this property holds. The set $\mathcal{A}$ is closed with respect to uniform limit passages and contains the set $\underline{P} - \underline{P}$ by Proposition 3.4.5. Therefore by the Deny–Hervé theorem [46: Theorem 2.3.1], $\mathcal{A}$ contains $\underline{\underline{\mathbb{C}}}_k(E)$, the subset of $\mathbb{C}_k(E)$ consisting of cylindrical functions. Consequently $\mathcal{A}$ contains $\mathbb{C}_k(E)$.

2) Let K be a compact set in E. For a function $f \in \mathbb{B}(E)$, we set

$$R_f^{E \setminus K} := \inf\{u \in \mathscr{S}^\mathbb{R} : u \geq f \cdot 1_{E \setminus K}\}.$$

We shall show that the following property holds:

$$\inf_{K \subset E} R_{R_0 1}^{E \setminus K} = 0.$$

We fix $\varepsilon > 0$ and choose N_ε such that $\forall \alpha > N_\varepsilon$:

$$R_0 1 \leq {}^\alpha R_0 1 + \varepsilon,$$

where ${}^\alpha R_0$ is the truncated kernel corresponding to the potential ${}^\alpha p = \sum_{\gamma \leq \alpha} p_\gamma$. We have ${}^\alpha p = {}^\alpha \overline{\overline{p}} \cdot \pi_\alpha$ (${}^\alpha \overline{\overline{p}} \in P_\alpha$) and ${}^\alpha R_0 1 = {}^\alpha p$. We write the chain of inequalities

$$\inf_{K \subset E} R_{R_0 1}^{E \setminus K} \leq \inf_{K \subset E} R_{{}^\alpha R_0 1 + \varepsilon}^{E \setminus K} \leq$$

$$\leq \inf_{K \subset E} R_{{}^\alpha R_0 1}^{E \setminus K} + \varepsilon \leq \inf_{K \subset E^\alpha} {}^\alpha R_{{}^\alpha \overline{\overline{p}}}^{E^\alpha \setminus K} \cdot \pi_\alpha + \varepsilon,$$

where ${}^\alpha R_{{}^\alpha \overline{\overline{p}}}^{E^\alpha \setminus K} = \inf\{u \in \mathscr{S}_\alpha^+(E^\alpha) : u \geq {}^\alpha \overline{\overline{p}} \cdot 1_{E^\alpha \setminus K}\}$. According to [46: Proposition 7.1.1]

$$\inf_{K \subset E^\alpha} {}^\alpha R_{{}^\alpha \overline{\overline{p}}}^{E^\alpha \setminus K} = 0,$$

therefore

$$\inf_{K \subset E} R_{R_0 1}^{E \setminus K} \leq \varepsilon.$$

In view of the arbitrariness of $\varepsilon > 0$, the required property is established.

3) We now list the properties of the resolvent $\mathbb{R}$:

(1) $R_\lambda : \mathbb{C}_b(E) \to \mathbb{C}_b(E)$ $(\forall \lambda \geq 0)$,

(2) $\lim_{\lambda \to \infty} \lambda R_\lambda \varphi(x) = \varphi(x)$ $(\forall \varphi \in \mathbb{C}_K(E), \ \forall x \in E)$,

(3) $\inf_{K \subset E} R_{R_0 1}^{E \setminus K} = 0$.

According to Taylor's theorem [104: Theorem 2.4, p. 82], every resolvent having these properties is generated by some standard process.[11] The proof is finished. $\square$

3.4.7

We shall fix a standard process $X = (x_t, \xi, \mathcal{M}_t, P^x)$ corresponding to resolvent $\mathbb{R}$ (Sec. 3.4.6) and consider the following objects:

$\{P_{E \setminus V}(x, \cdot)\}_{V \in \mathcal{U}, x \in V}$ is the family of hitting probabilities generated by process X,

$\{\mu_x^V\}_{V \in \mathcal{U}, x \in V}$ is the family of harmonic measures generated by a projective sequence of harmonic spaces $\{E^\alpha, \mathcal{S}_\alpha; \pi_{\alpha\beta}\}_1^\infty$ (Sec. 3.4.1), where $\underline{\underline{\mathcal{U}}}$ is the family of open cylindrical subsets of E.

Proposition 3.4.7. *The following property holds:*

$$\mu_x^V(\cdot) = P_{E \setminus V}(x, \cdot) \ (\forall V \in \underline{\underline{\mathcal{U}}}, \ \forall x \in V).$$

Proof. 1) We consider the set of functions[12] on E

$$\mathbb{S}_{\{\mathcal{S}_\alpha^+\}}^* := \{u \geq 0 : \mu_x^V(u) \leq u(x), \ \forall V \in \underline{\underline{\mathcal{U}}}, \ \forall x \in V\}$$

and show that the following inclusions hold:

$$\mathbb{S}_{\{\mathcal{S}_\alpha^+\}} \subset \mathcal{S}^{\mathbb{R}} \subset \mathbb{S}_{\{\mathcal{S}_\alpha^+\}}^*.$$

The inclusion $\mathbb{S}_{\{\mathcal{S}_\alpha^+\}} \subset \mathcal{S}^{\mathbb{R}}$ is proved in Sec. 3.4.5; we shall show that the inclusion $\mathcal{S}^{\mathbb{R}} \subset \mathbb{S}_{\{\mathcal{S}_\alpha^+\}}^*$ is satisfied. Since every function $u \in \mathcal{S}^{\mathbb{R}}$ is the limit of an increasing sequence of functions of the type $R_0 h$, $h \in \mathbb{B}^+(E)$, then it is sufficient to show that these functions are in $\mathbb{S}_{\{\mathcal{S}_\alpha^+\}}^*$. We set

$$\mathcal{A} := \{h \in \mathbb{B}^+(E) : R_0 h \in \mathbb{S}_{\{\mathcal{S}_\alpha^+\}}^*\}.$$

Let $h \in \underline{\mathbb{C}}_K^+(E)$; then $R_0 h = \sum_{\alpha=1}^\infty R_0^\alpha h$, where R_0^α is the kernel corresponding to potential p^α (see Proposition 3.4.4). But for each α we have $R_0^\alpha h \in \underline{\underline{P}} \subset \mathbb{S}_{\{\mathcal{S}_\alpha^+\}}^*$; consequently $R_0 h \in \mathbb{S}_{\{\mathcal{S}_\alpha^+\}}^*$. Thus $\underline{\mathbb{C}}_K^+(E) \subset \mathcal{A}$. The set $\mathcal{A}$ is closed with respect to

---

11 Even by a Hunt process (=standard process quasicontinuous from the left on $[0, \infty[$) [104: Corollary 2.8, p. 85].

12 In classical theory such functions are called *supermedian*; after regularization they become superharmonic (see e.g. [44]).

monotonic limit passages; therefore $\mathbb{C}_K^+(E) \subset \mathcal{A}$. Consequently $\mathcal{A}$ contains the set of all non-negative lower semicontinuous functions (we denote this set by $\mathcal{D}$), but then by the relation

$$R_0 h = \inf\{R_0 g : g \geq h, g \in \mathcal{D}\} \quad (\forall h \in \mathbb{B}^+(E))$$

$\mathcal{A}$ contains $\mathbb{B}^+(E)$.

2) We shall show that

$$\mu_x^V(u) = P_{E\backslash V}(x, u) \quad (\forall u \in \underline{\underline{P}}, \; \forall V \in \underline{\underline{\mathcal{U}}}, \; \forall x \in V).$$

We shall fix $u \in \underline{\underline{P}}$ and $V \in \underline{\underline{\mathcal{U}}}$ and show that

(a) $\mu_x^V(u) \geq P_{E\backslash V}(x, u) \quad (\forall x \in V).$

Let $u \in \underline{\underline{P}}_\alpha$ for some α , then on V we have

$$\mu_\cdot^V(u) = \inf_{\Gamma_\alpha} g, \quad \Gamma_\alpha = \{g \in \underline{\underline{\mathcal{S}}}_\alpha^+(E) : g \geq u \cdot 1_{E\backslash V}\},$$

$$P_{E\backslash V}(\cdot, u) = \inf_\Gamma g, \quad \Gamma = \{g \in \mathcal{S}^{\mathbb{R}} : g \geq u \cdot 1_{E\backslash V}\};$$

where the first relation is satisfied by [46: Proposition 5.5.3], and the second by Hunt's theorem [40: p. 141]. But by Proposition 3.4.5, $\Gamma_\alpha \subset \Gamma$, and therefore (a) is satisfied.

(b) We shall show that the inverse inequality is satisfied:

$$\mu_x^V(u) \leq P_{E\backslash V}(x, u) \quad (\forall x \in V).$$

First we note that since

$$P_{E\backslash V}(\cdot, u) = \inf\{P_{E\backslash W}(\cdot, u) : W \in \underline{\underline{\mathcal{U}}}, \; \overline{W} \subset V\},$$

then in the definition of Γ it is sufficient to consider functions $g \in \mathcal{S}^{\mathbb{R}}$ which dominate function u in some neighborhood of set $E\backslash V$ (its own neighborhood for each function g). Taking this into consideration, we choose $g \in \Gamma$ and set

$$\widetilde{g}(x) := \inf_{\pi_\alpha^{-1}(x)} g(y), \quad x \in E^\alpha,$$

$$\widetilde{\widetilde{g}}(x) := \sup_{W \downarrow \{x\}} \inf_W \widetilde{g},$$

$$g_\alpha := \widetilde{\widetilde{g}} \cdot \pi_\alpha.$$

The function $\widetilde{\widetilde{g}}$ is the lower-semicontinuous regularization of function $\widetilde{g}$, and consequently the function g_α on E is lower-semicontinuous. In addition,

$$g_\alpha \leq \widetilde{g} \cdot \pi_\alpha \leq g.$$

Let W be an open subset of E^α and let $^\alpha\mu^W$ be a harmonic measure of W (with respect to sheaf $\mathcal{S}_\alpha$); then $\forall x \in \pi_\alpha^{-1}(W)$ we have[13]

$$^\alpha\mu^W_{\pi_\alpha(x)}(\widetilde{\widetilde{g}}) = \mu_x^{\pi_\alpha^{-1}(W)}(g_\alpha) \leq \mu_x^{\pi_\alpha^{-1}(W)}(g) \leq g(x).$$

Consequently $\forall y \in W$

$$^\alpha\mu^W_y(\widetilde{\widetilde{g}}) \leq \widetilde{g}(y),$$

and since the left-hand side of the inequality is a continuous function on W, then after regularization we get

$$^\alpha\mu^W_y(\widetilde{\widetilde{g}}) \leq \widetilde{\widetilde{g}}(y) \quad (\forall y \in W).$$

By the arbitrariness of W and the axiom of completeness (for $\mathcal{S}_\alpha$) we conclude that $\widetilde{\widetilde{g}} \in \mathcal{S}_\alpha^+(E^\alpha)$, and consequently $g_\alpha \subset \underline{\mathcal{L}}_\alpha^+(E)$. In addition, since the set V and function u are cylindrical, and the function u is also continuous, then g_α dominates u in some neighborhood of $E\backslash V$. Consequently $g_\alpha \in \Gamma_\alpha$ and $g_\alpha \leq g$, and therefore

$$P_{E\backslash V}(x, u) = \inf_\Gamma g \geq \inf_\Gamma g_\alpha \geq$$
$$\geq \inf_{\Gamma_\alpha} g = \mu_x^V(u) \quad (\forall x \in V).$$

Thus the proof of inequality (b) is finished, which together with inequality (a) gives 2).

3) By 2) we have

$$\mu_x^V(u) = P_{E\backslash V}(x, u) \quad (\forall V \in \underline{\underline{\mathcal{U}}}, \ \forall x \in V, \ \forall u \in \underline{P} - \underline{P}).$$

By the Deny–Hervé theorem the noted equality is also satisfied for any function $\varphi \in \mathbb{C}_K(E)$, thus the proof of Proposition 3.4.7 is finished. $\square$

Corollary 3.4.1. *The process X has almost surely continuous trajectories.*

Proof. According to Proposition 3.4.7,

$$\mathrm{supp} P_{E\backslash V}(x, \cdot) \subset \partial V \quad (\forall V \subset \underline{\underline{\mathcal{U}}}, \ \forall x \in V).$$

Since $\underline{\underline{\mathcal{U}}}$ is the base of a topology of E, then by the theorem of Courrege and Priouret [47] the trajectories of X are almost surely continuous. $\square$

---

13 The inequality is satisfied by 1).

3.4.8

In this section we use the following notation:

$\mathscr{S}^{\mathbb{R}}$ is the cone of excessive functions with respect to resolvent $\mathbb{R}$;

$\mathscr{S}^{\mathbb{R}}_c \subset \mathscr{S}^{\mathbb{R}}$ is the cone of excessive lower-semicontinuous functions;

$\mathscr{S}^{\mathbb{R}}_\alpha = \{u \in \mathscr{S}^{\mathbb{R}} : u = \bar{\bar{u}} \cdot \pi_\alpha\},$

$\underline{\mathscr{S}}^+_\alpha = \{u = \bar{\bar{u}} \cdot \pi_\alpha : \bar{\bar{u}} \in \mathscr{S}^+_\alpha(E^\alpha)\},$

$\mathbb{S}_{\{\mathscr{S}^+_\alpha\}} = \{u = \lim_{\alpha \to \infty} u_\alpha : (\exists)\{u_\alpha\}^\infty_1, u_\alpha \in \underline{\mathscr{S}}^+_\alpha, u_\alpha \uparrow\}.$

Proposition 3.4.8. *The following relations hold:*
1) $\mathbb{S}_{\{\mathscr{S}^+_\alpha\}} = \mathscr{S}^{\mathbb{R}}_c,$
2) $\underline{\mathscr{S}}^+_\alpha = \mathscr{S}^{\mathbb{R}}_\alpha.$

Proof. 1) Let $u \in \mathscr{S}^{\mathbb{R}}_c$; for any α we set

$$\tilde{u}_\alpha(x) := \inf_{\pi_\alpha^{-1}(x)} u(y), \quad x \in E^\alpha,$$

$$\tilde{\tilde{u}}_\alpha(x) := \sup_{W \downarrow \{x\}} \inf_W \tilde{u}_\alpha,$$

$$u_\alpha := \tilde{\tilde{u}}_\alpha \cdot \pi_\alpha.$$

We have

$$u_\alpha \le \tilde{u}_\alpha \cdot \pi_\alpha \le u,$$

and in addition, as in the proof of Proposition 3.4.7 2) (b), we conclude that $u_\alpha \in \underline{\mathscr{S}}^+_\alpha$. We shall show that $u_\alpha \uparrow u$ for $\alpha \uparrow \infty$. In fact, let $\varphi \in \mathbb{C}^+_K(E) : \varphi \le u$. Then

$$\varphi_\alpha \le u_\alpha \le u.$$

Letting $\alpha \uparrow \infty$ and considering the continuity of φ and the fact that $u_\alpha \uparrow$, we obtain

$$\varphi \le \uparrow \lim_{\alpha \to \infty} u_\alpha \le u.$$

Passing to sup over all such φ in this inequality, we finally obtain

$$u \le \uparrow \lim_{\alpha \to \infty} u_\alpha \le u.$$

2) Let $u \in \mathscr{S}^{\mathbb{R}}_\alpha$, i.e. $u = \bar{\bar{u}} \cdot \pi_\alpha$; we assume also that the function u is bounded. Then by Proposition 3.4.4 (2) for any $\lambda \ge 0$ the function $R_\lambda u$ is continuous. Since $\lambda R_\lambda u \uparrow u, \lambda \uparrow \infty$, the function u is lower-semicontinuous and consequently is in $\mathscr{S}^{\mathbb{R}}_c$. According to 1)

$$u = \uparrow \lim_{\beta \to \infty} u_\beta, \quad u_\beta \in \underline{\mathscr{S}}^+_\beta \ (\forall \beta),$$

and now it remains to note that $u_\beta \equiv u_\alpha \ (\forall \beta \ge \alpha).$ $\qquad\square$

3.4.9

Let $\mathbb{P} = (P_t)_{t>0}$ be the Markov semigroup of a process X. In this subsection we establish the properties of this semigroup which are listed in Theorem 3.3.1. In the proof of these properties we shall follow the scheme of the proof of Theorem 10.2.2 of [46]. Since many details require modification in this connection, we shall give the proofs in their entirety.

Proposition 3.4.9. *The following property holds:*

$$P_t\varphi \in \mathbb{C}_b(E) \quad (\forall \varphi \in \mathbb{C}_K(E)).$$

Proof. We have $p = \sum_{\alpha=1}^{\infty} p_\alpha$. In correspondence with [46: Proposition 2.4.1], we choose a function $u \in \underline{P}_1$ such that $p_1 \leq \frac{1}{2}u$ and $p_1/u \in \mathbb{C}_0(E)$.[14] Then for any α the function u will be an Evans function of potential p_α, where $p_\alpha \leq 2^{-\alpha}p_1 \leq 2^{-(\alpha+1)}u$. Therefore the series $p/u = \sum_{\alpha=1}^{\infty} p_\alpha/u$ converges uniformly and consequently $p/u \in \mathbb{C}_0(E)$. We set

$$\mathscr{F} = \{f \in \mathbb{C}(E) : ||f||_u = \sup_E \frac{|f|}{u} < \infty\}.$$

It is obvious that $(\mathscr{F}, || \cdot ||_u)$ is a Banach space and $\mathbb{C}_K(E) \subset \mathscr{F}$. We set $\mathscr{F}_0 = \overline{\mathbb{C}_K(E)}$ and show that the following properties are satisfied:
(a) $R_\lambda : \mathscr{F}_0 \to \mathscr{F}_0, \forall \lambda \geq 0$,
(b) $\overline{R_\lambda(\mathscr{F}_0)} = \mathscr{F}_0, \forall \lambda \geq 0$.

These properties, according to the Hille–Yosida theorem, imply the existence of a semigroup $\mathbb{P} = (P_t)$ on $\mathscr{F}_0$ such that

$$R_\lambda = \int_0^\infty e^{-\lambda t}P_t\,dt \quad (\forall \lambda \geq 0).$$

It is obvious that $\mathbb{P}$ will coincide with the Markov semigroup of process X, and since

$$\mathscr{F}_0 = \{g = u \cdot v : v \in \mathbb{C}_0(E)\} \subset \mathbb{C}(E),$$

$$P_t(\mathbb{C}_K(E)) \subset \mathbb{B}(E) \cap P_t(\mathscr{F}_0) \subset \mathbb{B}(E) \cap \mathscr{F}_0 \subset \mathbb{C}_b(E),$$

then the proof of the proposition will be finished.

(a) We have $p/u \in \mathbb{C}_0(E)$, and consequently $p = p/u \cdot u \in \mathscr{F}_0$. Further, $\forall \varphi \in \mathbb{C}_K(E)$ we have

$$|\lambda R_\lambda\varphi| = |\lambda R_\lambda \frac{\varphi}{u} \cdot u| \leq ||\varphi||_u \lambda R_\lambda u \leq ||\varphi||_u \cdot u,$$

---

14 $\mathbb{C}_0(E)$ is the space of continuous functions tending to zero at ∞. The function u is called an Evans function of potential p_1 [46: p. 41].

consequently $\lambda R_\lambda \varphi \in \mathscr{F}$ and $||\lambda R_\lambda \varphi||_u \leq ||\varphi||_u$. In addition,

$$|R_\lambda \varphi| \leq \max |\varphi| R_0 1 = \max |\varphi| \cdot p,$$

consequently $R_\lambda \varphi = u \cdot R_\lambda \varphi / u \in \mathscr{F}_0$. Thus,

$$R_\lambda : \mathbb{C}_K(E) \to \mathscr{F}_0, \ ||\lambda R_\lambda||_u \leq 1,$$

which implies (a).

(b) Let $f \in \mathbb{C}_K(E^\alpha)$ and $\varepsilon > 0$. In correspondence with the Deny–Hervé theorem, we choose potentials $u_1, u_2 \in P_\alpha$ with compact harmonic supports $S(u_1)$ and $S(u_2)$ such that

$$|f - (u_1 - u_2)| < \frac{\varepsilon}{2} \overline{\overline{u}} \cdot \pi_{1\alpha}$$

where $u = \overline{\overline{u}} \cdot \pi_1$. For some number $c > 0$ we will have $(i = 1, 2)$

$$u_i(x) \leq c \overline{\overline{p}}_\alpha(x), \ \forall x \in S(u_i), \ p_\alpha = \overline{\overline{p}}_\alpha \cdot \pi_\alpha,$$

and by the domination principle [2: p. 31] this inequality will be satisfied $\forall x \in E^\alpha$. Consequently

$$u_i \cdot \pi_\alpha(x) \leq cp(x), \ \forall x \in E \ (i = 1, 2).$$

Therefore $u_i \cdot \pi_\alpha / u \in \mathbb{C}_0(E)$, and therefore $u_i \cdot \pi_\alpha \in \mathscr{F}_0$. Further, $\lambda R_\lambda(u_i \cdot \pi_\alpha) \uparrow u_i \cdot \pi_\alpha$, $\lambda \uparrow \infty$, and consequently by Dini's theorem there exists a $\nu > 0$ such that

$$u_i \cdot \pi_\alpha(x) \leq \frac{\varepsilon}{4} u + \nu R_\nu(u_i \cdot \pi_\alpha), \ \forall x \in \pi_\alpha^{-1}(S(u_i)).$$

But then $\forall \beta > \alpha$

$$u_i \cdot \pi_{\alpha\beta}(x) \leq \left[\frac{\varepsilon}{4} u + \nu R_\nu(u_i \cdot \pi_\alpha)\right]_\beta (x), \ \forall x \in \pi_{\alpha\beta}^{-1}(S(u_i)),$$

where $[....]_\beta(x) = \inf_{\pi_\beta^{-1}(x)}[....]$. We note that since $[....] \in \mathbb{C}(E)$, then also $[....]_\beta \in \mathbb{C}(E^\beta)$. We have $u_i \cdot \pi_{\alpha\beta} \in P_\beta$, $u \in \underline{P}_1 \subset \mathscr{S}_c^{\mathbb{R}}$ and

$$R_\nu(u_i \cdot \pi_\alpha) = R_0(u_i \cdot \pi_\alpha - \nu R_\nu(u_i \cdot \pi_\alpha)) \in \mathscr{S}_c^{\mathbb{R}},$$

consequently also $[....] \in \mathscr{S}_c^{\mathbb{R}}$, and therefore $[....]_\beta \in \mathscr{S}_\beta^+(E^\beta)$. By the domination principle the above-noted inequality is satisfied everywhere on E^β; letting $\beta \uparrow \infty$ we get

$$u_i \cdot \pi_\alpha \leq \frac{\varepsilon}{4} u + \nu R_\nu(u_i \cdot \pi_\alpha).$$

Now we write the chain of inequalities

$$|f \cdot \pi_\alpha - R_\nu(\nu(u_1 - u_2)\pi_\alpha)| \leq$$

$$\leq |f \cdot \pi_\alpha - (u_1 - u_2) \cdot \pi_\alpha| + |u_1 \cdot \pi_\alpha - \nu R_\nu(u_1 \cdot \pi_\alpha)| +$$

$$+ |u_2 \cdot \pi_\alpha - \nu R_\nu (u_2 \cdot \pi_\alpha)| \leq \varepsilon u.$$

Thus, setting $g = \nu(u_1 - u_2) \cdot \pi_\alpha$, we shall have

$$\|f \cdot \pi_\alpha - R_\nu g\|_u < \varepsilon,$$

and in addition,

$$R_\nu g = R_\lambda[g + (\lambda - \nu)R_\nu g] := R_\lambda d.$$

Since $g \in \mathcal{F}_0$ and $R_\nu g \in \mathcal{F}_0$, then also $d \in \mathcal{F}_0$. Thus for the given $f \in \mathbb{C}_k(E^\alpha)$ and $\varepsilon > 0$, we can find $d \in \mathcal{F}_0$ such that

$$\|f \cdot \pi_\alpha - R_\lambda d\|_u < \varepsilon,$$

consequently $\overline{R_\lambda(\mathcal{F}_0)}$ contains all cylindrical functions of $\mathbb{C}_k(E)$, and therefore also $\mathbb{C}_k(E)$. Thus property (b) is proved. $\qquad\square$

Up to now we have considered the set $\mathcal{S}^\mathbb{R}$ of excessive functions with respect to resolvent $\mathbb{R} = (R_\lambda)_{\lambda \geq 0}$, i.e. the set of functions $u \geq 0$ such that

$$(\forall \lambda > 0) \; \lambda R_\lambda u \leq u, \; \lambda R_\lambda u \uparrow u, \; \lambda \uparrow \infty.$$

According to [81: Theorem 65, p. 241] the set of such functions coincides with the set $\mathcal{S}^\mathbb{P}$ of excessive functions with respect to a semigroup $\mathbb{P} = (P_t)_{t>0}$, i.e. functions $u \geq 0$ such that

$$(\forall t > 0) \; P_t u \leq u, \; P_t u \uparrow u, \; t \downarrow 0.$$

Based on this remark, we shall prove the following proposition.

Proposition 3.4.10. *The following property holds:*

$$\lim_{t \downarrow 0} \sup_E |P_t \varphi - \varphi| = 0, \; \forall \varphi \in \mathbb{C}_k(E).$$

Proof. We choose $q \in P_\alpha$ such that
(a) $S(q)$ is a compact set in E^α,
(b) $q \leq \overline{\overline{p}}_\alpha \; (p_\alpha = \overline{\overline{p}}_\alpha \cdot \pi_\alpha)$.
According to (b), $q \cdot \pi_\alpha < p$, and consequently $q \cdot \pi_\alpha \in \mathcal{F}_0$. Therefore we conclude that

$$(\forall t > 0) \; P_t(q \cdot \pi_\alpha) \in \mathcal{F}_0 \cap \mathbb{B}(E) \subset \mathbb{C}_b(E).$$

On the other hand, $q \cdot \pi_\alpha \in \underline{\underline{P}} \subset \mathcal{S}^\mathbb{P}$ and therefore $P_t(q \cdot \pi_\alpha) \uparrow q \cdot \pi_\alpha$ for $t \uparrow 0$. By Dini's theorem

$$\lim_{t \downarrow 0} \sup_{\pi_\alpha^{-1}(S(q))} (q \cdot \pi_\alpha - P_t(q \cdot \pi_\alpha)) = 0.$$

We write the obvious inequality: $\forall t > 0$, $\forall x \in \pi_\alpha^{-1}(S(q))$,

$$q \cdot \pi_\alpha(x) \le P_t(q \cdot \pi_\alpha)(x) + \sup_{\pi_\alpha^{-1}(S(q))} (q \cdot \pi_\alpha - P_t(q \cdot \pi_\alpha)).$$

Passing to $\inf_{\pi_\beta^{-1}(\cdot)}$, $\beta > \alpha$, on both sides of this inequality we obtain: $\forall t > 0$, $\forall x \in \pi_{\alpha\beta}^{-1}(S(q))$

$$q \cdot \pi_{\alpha\beta}(x) \le [P_t(q \cdot \pi_\alpha)]_\beta(x) + \sup_{\pi_\alpha^{-1}(S(q))} (q \cdot \pi_\alpha - P_t(q \cdot \pi_\alpha))$$

where $[P_t(q \cdot \pi_\alpha)]_\beta(x) = \inf_{\pi_\beta^{-1}(x)}[P_t(q \cdot \pi_\alpha]$. We note further that $q \cdot \pi_{\alpha\beta} \in P_\beta$ and $[P_t(q \cdot \pi_\alpha)]_\beta \in P_\beta$, and in addition $S(q \cdot \pi_{\alpha\beta}) \subset \pi_{\alpha\beta}^{-1}(S(q))$. According to the domination principle in $(E^\beta, \mathscr{S}_\beta)$, the above-noted inequality will be satisfied $\forall x \in E^\beta$. Now passing to the limit for $\beta \uparrow \infty$ on both sides of this inequality, and considering the continuity of the functions $q \cdot \pi_\alpha$ and $P_t(q \cdot \pi_\alpha)$, we get

$$(\forall t > 0) \; q \cdot \pi_\alpha \le P_t(q \cdot \pi_\alpha) + \sup_{\pi_\alpha^{-1}(S(q))} (q \cdot \pi_\alpha - P_t(q \cdot \pi_\alpha)).$$

Finally, from this inequality we get

$$\limsup_{t \downarrow 0} {}_E (q \cdot \pi_\alpha - P_t(q \cdot \pi_\alpha)) \le$$

$$\le \lim_{t \downarrow 0} \sup_{\pi_\alpha^{-1}(S(q))} (q \cdot \pi_\alpha - P_t(q \cdot \pi_\alpha)) = 0.$$

Now let $\overline{\overline{\varphi}} \in \mathbb{C}_k^+(E^\alpha)$ and $\varphi = \overline{\overline{\varphi}} \cdot \pi_\alpha$. Since the potential $\overline{p}_\alpha$ is hyperstrict, for a given $\varepsilon > 0$ it is possible to choose potentials q' and q'' that have properties (a) and (b), and in addition

$$0 \le q' - q'' \le \overline{\overline{\varphi}} \le q' - q'' + \varepsilon/2.$$

But then after the obvious transformations we get

$$\limsup_{t \downarrow 0} \sup_E |\varphi - P_t\varphi| \le \varepsilon.$$

In view of the arbitrariness of $\varepsilon > 0$,

$$\limsup_{t \downarrow 0} {}_E | \varphi - P_t\varphi | = 0.$$

Now it remains to note that the set of cylindrical functions in $\mathbb{C}_k(E)$ is dense in $\mathbb{C}_k(E)$. The proof is complete. $\qquad\square$

Proposition 3.4.11. *For any function $\varphi \in \mathbb{C}_k(E)$ and any closed set $F : F \cap \operatorname{supp} \varphi = \emptyset$, the following equality holds:*

$$\lim_{t \downarrow 0} \frac{1}{t} \sup_F P_t|\varphi| = 0.$$

Proof. Let a function φ and a set F satisfy the conditions of the proposition. For some α we shall choose a function $f \in \mathbb{C}_k^+(E^\alpha)$ such that:
(a) supp $(f \cdot \pi_\alpha) \cap F = \emptyset$,
(b) $|\varphi| \le f \cdot \pi_\alpha$.

It is obviously sufficient to show that the following equality holds:

$$\lim_{t \downarrow 0} \frac{1}{t} \sup_F P_t |f \cdot \pi_\alpha| = 0. \tag{$*$}$$

We choose p', $p'' \in P_\alpha$ such that the following conditions are satisfied:
(a) $p' + p'' \le \delta \overline{\overline{p}}_\alpha$ $(\exists \delta > 0)$ $(p_\alpha = \overline{\overline{p}}_\alpha \cdot \pi_\alpha)$,
(b) $S(p') \cup S(p'') \subset E^\alpha \setminus \pi_\alpha(F)$,
(c) $f \le p'' - p'$,
(d) $p' = p''$ on the set $\pi_\alpha(F)$.

We have on F:

$$\frac{1}{t} P_t(f \cdot \pi_\alpha) \; \le \; \frac{1}{t} P_t(p'' \cdot \pi_\alpha - p' \cdot \pi_\alpha) =$$

$$= \frac{1}{t}[(p' \cdot \pi_\alpha - p'' \cdot \pi_\alpha) - P_t(p' \cdot \pi_\alpha - p'' \cdot \pi_\alpha)] \le$$

$$\le \frac{1}{t}[p' \cdot \pi_\alpha - P_t(p' \cdot \pi_\alpha)].$$

Thus, the equality $(*)$ will be established if we can show that $\forall u \in P_\alpha$:
(a) $u \le \delta \overline{\overline{p}}_\alpha$ $(\exists \delta > 0)$,
(b) $S(u)$ is a compact set in E^α,

the following equality

$$\lim_{t \downarrow 0} \frac{1}{t} \sup_{E \setminus V}(u \cdot \pi_\alpha - P_t(u \cdot \pi_\alpha)) = 0 \tag{$**$}$$

is satisfied, where V is a neighborhood of the compact set $\pi_\alpha^{-1}(S(u))$ in E.

Let $\mathbb{R}^\alpha = (R_\lambda^\alpha)_{\lambda \ge 0}$ be the Feller resolvent on E associated with potential p_α (see Proposition 3.4.3) and also let $\overline{\overline{\mathbb{R}}}^\alpha = (\overline{\overline{R}}_\lambda^\alpha)_{\lambda \ge 0}$ be the resolvent on E^α associated with the potential $\overline{\overline{p}}_\alpha : p_\alpha = \overline{\overline{p}}_\alpha \cdot \pi_\alpha$. We have

$$R_\lambda^\alpha(\varphi \cdot \pi_\alpha) = (\overline{\overline{R}}_\lambda^\alpha \varphi) \cdot \pi_\alpha \; (\forall \lambda \ge 0, \; \forall \varphi \in \mathbb{B}(E^\alpha)).$$

From the fact that $u \le \delta \overline{\overline{p}}_\alpha$, it follows (see [46: Ex. 10.2.1]) that there exists a function $g \in \mathbb{B}(E^\alpha)$ such that
(a) $0 \le g \le \delta$,
(b) $u = \overline{\overline{R}}_0^\alpha g$,
(c) supp $g = S(u)$.

But then

$$u \cdot \pi_\alpha = (\overline{\overline{R}}_0^\alpha g) \cdot \pi_\alpha = R_0^\alpha(g \cdot \pi_\alpha).$$

Now let $\mathbb{R} = (R_\lambda)_{\lambda \geq 0}$ be a resolvent of semigroup $\mathbb{P} = (P_t)$; then

$$R_0(g \cdot \pi_\alpha) = \sum_{\beta=1}^{\infty} R_0^\beta(g \cdot \pi_\alpha) =$$

$$= R_0^\alpha(g \cdot \pi_\alpha) + \sum_{\beta \neq \alpha} R_0^\beta(g \cdot \pi_\alpha) = u \cdot \pi_\alpha + \mathcal{V},$$

where $\mathcal{V} = \sum_{\beta \neq \alpha} R_0^\beta(g \cdot \pi_\alpha) \in \mathscr{S}^{\mathbb{P}}$.

From this we get:

$$\frac{1}{t}(u \cdot \pi_\alpha - P_t(u \cdot \pi_\alpha)) \leq \frac{1}{t}(R_0(g \cdot \pi_\alpha) - P_t(R_0(g \cdot \pi_\alpha))) =$$

$$= \frac{1}{t}\int_0^t P_s(g \cdot \pi_\alpha)\, ds \leq \frac{1}{t}\int_0^t P_s \tilde{g}\, ds,$$

where $\tilde{g}$ is a function of $\mathbb{C}_k^+(E)$ such that:

(a) $(g \cdot \pi_\alpha) \leq \tilde{g}$,

(b) $\operatorname{supp} \tilde{g} \subset V$.

Now applying Proposition 3.4.10 we finally get:

$$\limsup_{\substack{t \downarrow 0 \\ E \setminus V}} \frac{1}{t}(u \cdot \pi_\alpha - P_t(u \cdot \pi_\alpha)) \leq$$

$$\leq \limsup_{\substack{t \downarrow 0 \\ E \setminus V}} \frac{1}{t}\int_0^t P_s \tilde{g}\, ds =$$

$$= \limsup_{\substack{t \downarrow 0 \\ E \setminus V}} \frac{1}{t}\int_0^t [P_s \tilde{g} - \tilde{g}]\, ds \leq$$

$$\leq \lim_{t \downarrow 0} \frac{1}{t}\int_0^t \sup_E |P_s \tilde{g} - \tilde{g}|\, ds = 0.$$

Thus equation (**) is proved, together with Proposition 3.4.11.   □

3.4.10

Proof of Theorem 3.3.1. We choose a sequence $\{p_\alpha\}_1^\infty \subset \underline{P}$ such that the following conditions are satisfied:

1) $p_\alpha = \overline{\overline{p}}_\alpha \cdot \pi_\alpha, \overline{\overline{p}}_\alpha \in P_\alpha$;

2) $\overline{\overline{p}}_\alpha$ is a hyperstrict potential on $(E^\alpha, \mathscr{S}_\alpha)$;

3) $p_{\alpha+1} \leq \frac{1}{2}p_\alpha, p_1 \leq \frac{1}{2}$,

and set $p = \sum_{\alpha=1}^\infty p_\alpha$. According to Proposition 3.4.4, there exists a resolvent $\mathbb{R} = (R_\lambda)_{\lambda \geq 0}$ on $\mathbb{B}(E)$ such that:

1) $R_0 1 = p$,

2) $R_\lambda : \mathbb{C}_b(E) \cup \underline{\mathbb{B}}(E) \to \mathbb{C}_b(E)$ $(\forall \lambda \geq 0)$.

According to Proposition 3.4.6, there exists a standard process $X = (x_t, \xi, \mathcal{M}_t, P^X)$ in phase space $(E, \mathscr{B})$ such that

$$R_\lambda \varphi(x) = M^x \left(\int_0^\xi e^{-\lambda t} \varphi(x_t) \, dt \right) \quad (\forall \lambda \geq 0, \ \forall \varphi \in \mathbb{B}(E));$$

by Corollary 3.4.1 (Sec. 3.4.7) this process has almost surely continuous trajectories. According to Propositions 3.4.9-3.4.11, the Markov semigroup $\mathbb{P} = (P_t)_{t>0}$ of process X satisfies the conditions:

1. $\forall \varphi \in \mathbb{C}_k(E)$:
   (a) $P_t \varphi \in \mathbb{C}_b(E)$ $(\forall t > 0)$,
   (b) $\lim_{t \downarrow 0} \sup_E |P_t \varphi - \varphi| = 0$,
   (c) $\forall F \subset E : \overline{F} \cap \operatorname{supp} \varphi = \emptyset$, we have

$$\lim_{t \downarrow 0} \frac{1}{t} \sup_F P_t |\varphi| = 0.$$

And finally, according to Proposition 3.4.8, the following conditions are satisfied:

2.   (a) $\mathscr{S}_\alpha^{\mathbb{P}} = \underline{\mathscr{L}}_\alpha^+$ $(\forall \alpha)$,
   (b) $\mathscr{S}_c^{\mathbb{P}} = \mathbb{S}_{\{\mathscr{S}_\alpha^+\}}$.

3.4.11

Proof of Theorem 3.3.2. Let $\mathscr{S}^X$ be the sheaf of locally superharmonic functions of process X and let $\mathscr{S}$ be the subsheaf of $\mathscr{S}^X$ consisting of lower-semicontinuous functions. For every α, by Proposition 3.4.7 and Sec. 3.4.1, the inclusion

$$\mathscr{S}_\alpha \cdot \pi_\alpha \subset \mathscr{S}$$

holds (see Sec. 3.3.2); in particular

$$\mathscr{H}_{\mathscr{S}_\alpha} \cdot \pi_\alpha \subset \mathscr{H}_{\mathscr{S}}.$$

Since each of the sheaves $\mathscr{H}_{\mathscr{S}_\alpha}$ is non-degenerate, the sheaf $\mathscr{H}_{\mathscr{S}}$ is also non-degenerate.

It is obvious that $\mathscr{H}_{\mathscr{S}}$ is closed with respect to uniform limit passages.

We consider the set $V \in \mathfrak{U}_k$ in a neighborhood W of which there exists a strictly positive harmonic function $\overline{h} \in \mathscr{H}_{\mathscr{S}}^+(W)$. We shall show that V is a resolutive set. From the choice of V and [54: Sec. 1.2.16] it follows that $\mathscr{S}(V)$ coincides with the set of lower-semicontinuous superharmonic functions for process X on set V. Let the function $f \geq 0$ be continuous and belong to $\underline{P} = \cup_\alpha \underline{P}_\alpha$ and let $u \in \mathscr{S}(V)$ such that

$$\lim_{x \to y, \, x \in V} \inf u(x) \geq f(y) \quad (\forall y \in \partial V).$$

We choose a sequence $\{V_n\}_1^\infty \subset \mathcal{U}_k$ such that $\overline{V}_n \subset V_{n+1}$ and $V = \cup_{n \geq 1} V_n$; then

$$u(x) \geq \liminf_{n \to \infty} P_{E \backslash V_n}(x, u) = \liminf_{n \to \infty} M_x[u(x_{T_{E \backslash V_n}})] \geq$$

$$\geq M_x[\liminf_{n \to \infty} u(x_{T_{E \backslash V_n}})] \geq M_x[f(x_{T_{E \backslash V}})] = P_{E \backslash V}(x, f),$$

from which it follows that, first, V is an MP-set, and second, that $\overline{H}_f \geq P_{E \backslash V}(\cdot, f)$.[15] On the other hand, by Hunt's theorem [40: Theorem 6.12] the function $P_{E \backslash V}(\cdot, f)$ coincides on V with the lower envelop of the family of excessive functions which majorize the function f on a neighborhood (distinct for each function) of set $E \backslash V$. Thus, as in the proof of Proposition 3.4.7, we deduce that $P_{E \backslash V}(\cdot, f) \geq \overline{H}_f$. Consequently, $\overline{H}_f = P_{E \backslash V}(\cdot, f)$. Further, from Proposition 3.4.7 it follows that the function $P_{E \backslash V}(\cdot, f)$ is continuous and harmonic on V. Therefore $P_{E \backslash V}(\cdot, f) \in (-\mathcal{S})(V)$. This property together with the inequality $P_{E \backslash V}(\cdot, f) \leq f$ shows that $P_{E \backslash V}(\cdot, f) \leq \underline{H}_f$. Finally we get

$$P_{E \backslash V}(\cdot, f) \leq \underline{H}_f \leq \overline{H}_f = P_{E \backslash V}(\cdot, f),$$

and consequently the function f is resolutive. The set of resolutive functions is a linear space [46: p. 19], therefore every function $f \in \underline{P} - \underline{P}$ is resolutive. The resolutivity of any function $f \in \mathbb{C}_k(E)$ now follows from the possibility of uniform approximation on ∂V of function f by functions from the set $\underline{P} - \underline{P}$.

We note that in passing we have established the equality

$$H_f(x) = P_{E \backslash V}(x, f) \quad (\forall x \in V).$$

Thus, the hitting probability $P_{E \backslash V}(x, \cdot)$ exactly coincides with harmonic measure $\mu_x^V(\cdot)$ of set V (see Sec. 2.2). From this equality, in turn, it follows that in the space $(E, \mathcal{S})$ the axiom of completeness is satisfied. The axiom of existence of a potential is satisfied as well, since for any potential $\overline{\overline{p}}$ with respect to a sheaf $\mathcal{S}_\alpha$, the function $p = \overline{\overline{p}} \cdot \pi_\alpha$ is a potential with respect to sheaf $\mathcal{S}$. $\square$

3.4.12

Proof of Proposition 3.3.2. Let $\{X^\alpha; \pi_{\alpha\beta}\}_1^\infty$ be a projective sequence of Markov processes satisfying the conditions of Proposition 3.3.2, and let $\{\mathcal{S}_\alpha\}_1^\infty$ be its corresponding sequence of hyperharmonic sheaves: $\mathcal{S}_\alpha = \mathcal{S}^{X^\alpha}$. We fix $\alpha < \beta$ and show that

1) $\mathcal{S}_\alpha \cdot \pi_{\alpha\beta} \subset \mathcal{S}_\beta$.

In view of the arbitrariness of the choice of the pair $\alpha < \beta$, this also means that the sequence of harmonic spaces $\{E^\alpha, \mathcal{S}_\alpha; \pi_{\alpha\beta}\}_1^\infty$ is projective.

From the relation $X^\alpha = \pi_{\alpha\beta}(X^\beta)$ it immediately follows that if the function f is excessive with respect to process X^α, then the function $f \cdot \pi_{\alpha\beta}$ is excessive with

---

15 $\overline{H}_f$ and $\underline{H}_f$ are the upper and lower solutions of the generalized Dirichlet problem (see Sec. 2.2).

respect to process X^β; thus we have

$$\mathcal{S}^+_\alpha(E^\alpha) \cdot \pi_{\alpha\beta} \subset \mathcal{S}^+_\beta(E^\beta).$$

Now let V^α be an open subset of E^α and let $V^\beta := \pi^{-1}_{\alpha\beta}(V^\alpha)$. According to Theorem 3.3.3, the trajectories x^α_t and x^β_t of processes X^α and X^β are connected by the relation

$$x^\alpha_t = \pi_{\alpha\beta}(x^\beta_t),$$

therefore

$$T_{E^\alpha \setminus V^\alpha} = T_{E^\beta \setminus V^\beta},$$

where $T_{E^\alpha \setminus V^\alpha}$ and $T_{E^\beta \setminus V^\beta}$ are the first hitting times of processes X^α and X^β from sets V^α and V^β respectively. From this we get

2) $X^\alpha_{V^\alpha} = \pi_{\alpha\beta}(X^\beta_{V^\beta})$,

3) $P_{E^\alpha \setminus V^\alpha}(\pi_{\alpha\beta}(x), \Gamma) = P_{E^\beta \setminus V^\beta}(x, \pi^{-1}_{\alpha\beta}(\Gamma))$

where $X^\alpha_{V^\alpha}$ and $X^\beta_{V^\beta}$ are the parts of processes X^α and X^β on sets V^α and V^β respectively, and $P_{E^\alpha \setminus V^\alpha}$ and $P_{E^\beta \setminus V^\beta}$ are the corresponding hitting probabilities. Since the set $\mathcal{S}^+_\alpha(V_\alpha)$ coincides with the set of excessive functions with respect to process $X^\alpha_{V^\alpha}$, then from relation 2) it follows that

4) $\mathcal{S}^+_\alpha(V^\alpha) \cdot \pi_{\alpha\beta} \subset \mathcal{S}^+_\beta(V^\beta)$,

and from relation 3) it follows that

5) $\mathcal{H}_{\mathcal{S}_\alpha}(V^\alpha) \cdot \pi_{\alpha\beta} \subset \mathcal{H}_{\mathcal{S}_\beta}(V^\beta)$.

Now let the set V^α be such that in a neighborhood of it there exists a strictly positive harmonic function $h \in \mathcal{H}^+_{\mathcal{S}_\alpha}(V^\alpha)$. Let $f \in \mathcal{S}_\alpha(V^\alpha)$, then for some number $c > 0$ we have $f + ch \in \mathcal{S}^+_\alpha(V^\alpha)$ and according to 4), we have $f \cdot \pi_{\alpha\beta} + ch \cdot \pi_{\alpha\beta} \in \mathcal{S}^+_\beta(V^\beta)$. Therefore by 5) we have $f \cdot \pi_{\alpha\beta} \in \mathcal{S}_\beta(V^\beta)$. Thus it is established that

6) $\mathcal{S}_\alpha(V^\alpha) \cdot \pi_{\alpha\beta} \subset \mathcal{S}_\beta(V^\beta)$

for the specific set V^α. Now we note that the family of these sets forms the base of a topology, therefore in the general case the function $f \cdot \pi_{\alpha\beta}$ will be in $\mathcal{S}_\beta(W)$, where W ranges over an open cover of set V^β. Since $\mathcal{S}_\beta$ is a sheaf, it will follow that $f \cdot \pi_{\alpha\beta} \in \mathcal{S}_\beta(V^\beta)$. Thus inclusion 6) holds for any open set V^α, which proves 1).

Now let $X = \lim \mathrm{proj}\, X^\alpha$ and $\mathcal{S} := \mathcal{S}^X_c$ be the sheaf of lower-semicontinuous locally superharmonic functions of process X. We have

$$X^\alpha = \pi_\alpha(X) \quad (\forall \alpha),$$

therefore arguing as in the case of the pair X^α and X^β, we arrive at the inclusion

$$\mathcal{S}_\alpha \cdot \pi_\alpha \subset \mathcal{S} \quad (\forall \alpha).$$

Now it remains to show that the sheaf $\mathcal{S}$ is minimal (see Sec. 3.3.2). Let $V = \pi^{-1}_\alpha(V^\alpha)$ be an open cylindrical subset of E and let $f \in \mathcal{S}(V)$. For any $\beta < \alpha$ we set

$$\tilde{f}_\beta(x) := \inf_{\pi_\beta^{-1}(x)} f(y), \quad x \in V^\beta = \pi_{\alpha\beta}^{-1}(V^\alpha),$$

$$\widetilde{\tilde{f}}_\beta(x) := \sup_{W\downarrow\{x\}} \inf_W \tilde{f},$$

$$f_\beta := \widetilde{\tilde{f}}_\beta \cdot \pi_\beta.$$

As in part 1) of the proof of Proposition 3.4.8, we deduce that

a) $\widetilde{\tilde{f}}_\beta \in \mathscr{S}_\beta(V^\beta)$,

b) $f_\beta \uparrow f, \beta \uparrow \infty$.

It is obvious that these properties imply the minimality of sheaf $\mathscr{S}$. $\square$

3.5 Projective sequences of harmonic spaces: some remarks on harmonic functions on a Wiener space

Now we discuss Example 3.3.1. This example is not included in the general scheme of Sec. 3.3 because the mappings

$$\pi_{\alpha\beta} : \mathbb{R}^\beta \to \mathbb{R}^\alpha \quad (\alpha \le \beta)$$

do not satisfy condition 4) (Sec. 3.3.3). Consequently the limit space

$$E = \lim\operatorname{proj}\mathbb{R}^\alpha = \mathbb{R}^\infty$$

is not locally compact and therefore we automatically leave the framework of general potential theory [46], [39]. Construction of a potential theory on a non-locally-compact space, as noted in [60], meets with some difficulties, including some connected with the peculiarities of measure theory on such a space [100]. We present here some results from [60] and show that Example 3.3.1 naturally fits in with the theory constructed there. Along with this we note the connection between Example 3.3.1 and Examples 3.3.2 and 3.3.3.

3.5.1

We shall carry out the discussion in a somewhat more general situation than in Example 3.3.1. In fact, we shall fix a sequence of positive numbers $\{a_k\}_1^\infty$ and $\forall \alpha = 1, 2, \ldots$ we set

$$\mathscr{L}_\alpha := \sum_{k=1}^{\alpha} a_k \partial_k^2, \quad \mathscr{S}_\alpha := \mathscr{S}_{\mathscr{L}_\alpha}.$$

We have

1) $(\mathbb{R}^\alpha, \mathcal{S}_\alpha)$ is a $\mathcal{P}$-harmonic space $(\alpha \geq 3)$;

2) $\mathcal{S}_\alpha \cdot \pi_{\alpha\beta} \subset \mathcal{S}_\beta$ $(\forall \alpha, \beta : \alpha < \beta)$.

Later constructions are carried out by the same scheme as in Sec. 3.3. For every α we shall find a Markov process X_α on phase space $(\mathbb{R}^\alpha, \mathcal{B}(\mathbb{R}^\alpha))$ such that, first, sheaf $\mathcal{S}_\alpha$ coincides with the sheaf of superharmonic functions of process X_α, and second, the sequence of processes $\{X_\alpha\}_1^\infty$ is projective with respect to the system of mappings $\{\pi_{\alpha\beta}\}_1^\infty$. Further, following the method of Sec. 3.3.4, one should consider the process $X_\infty = \lim \operatorname{proj} X_\alpha$ and choose the sheaf of its superharmonic functions as a limit sheaf. However, here we avoid the projective limit scheme for the following reason. It turns out that there exists a Banach space $B \subset \mathbb{R}^\infty$ having the property that trajectories of process X_∞ exiting at the initial time from B, with probability one do not leave B. Therefore for the choice of a "limit" sheaf, it is expedient to start not with the process X_∞ but with the process obtained from the restriction of X_∞ to B.

Let n_t be a Gaussian measure on $(\mathbb{R}^1, \mathcal{B}(\mathbb{R}^1))$ with parameters $(0, 2t)$. We define measures

$$^\alpha\mu_t = \otimes_{k=1}^\alpha n_{a_k t},$$
$$\mu_t = \otimes_{k=1}^\infty n_{a_k t},$$

on $(\mathbb{R}^\alpha, \mathcal{B}(\mathbb{R}^\alpha))$ and $(\mathbb{R}^\infty, \mathcal{B}(\mathbb{R}^\infty))$, and we denote by X_α and X_∞ the strict Markov processes with almost surely continuous trajectories on $\mathbb{R}^\alpha$ and $\mathbb{R}^\infty$ respectively, which are defined by the transition functions

$$^\alpha p(t; x, \Gamma) = {}^\alpha\mu_t(\Gamma - x), \ \Gamma \in \mathcal{B}(\mathbb{R}^\alpha),$$
$$p(t; x, \Gamma) = \mu_t(\Gamma - x), \ \Gamma \in \mathcal{B}(\mathbb{R}^\infty).$$

We give two properties of these processes, the first of which follows from [54: Chapter 14, 1.], and the second from Theorem 3.3.3.

3) The sheaf $\mathcal{S}_\alpha$ coincides with the sheaf of superharmonic functions of process X_α $(\forall \alpha \geq 1)$;

4) $X_\alpha = \pi_{\alpha\beta}(X_\beta)$ $(\forall \alpha, \beta : \alpha < \beta \leq \infty)$.

Thus the sequence $\{X_\alpha; \pi_{\alpha\beta}\}_1^\infty$ is projective and X_∞ is its projective limit. We consider the sets

$$H = \left\{ x = (x_k)_1^\infty \in \mathbb{R}^\infty : |x|^2 := \sum_{k=1}^\infty x_k^2/a_k < \infty \right\},$$

$$B = \left\{ x = (x_k)_1^\infty \in \mathbb{R}^\infty : ||x||^2 := \sum_{k=1}^\infty x_k^2/2^k(1 + a_k) < \infty \right\}.$$

It is not difficult to see that H and B are Hilbert subspaces of $\mathbb{R}^\infty$, $H \subset B$, and in addition, the space B coincides with the completion of space H with respect to norm $\|\cdot\|$.

For each α and r we set

$$C_{\alpha,r} := \left\{ x \in \mathbb{R}^\infty : \sum_{k=1}^{\alpha} x_k^2/2^k(1+a_k) \geq r \right\}$$

and we apply the well-known inequality for Gaussian measures [49: Lemma 1.1, p. 58]:

$$\mu_t(\mathbb{R}^\infty \setminus B) = \mu_t(\cap_r \cup_\alpha C_{\alpha,r}) = \inf_r \sup_\alpha \mu_t(C_{\alpha,r}) =$$

$$\inf_r \sup_\alpha {}^\alpha\mu_t(\pi_\alpha(C_{\alpha,r})) \leq \inf_r \sup_\alpha \frac{2t}{r} \sum_{k=1}^{\alpha} \frac{1}{2^k} = 0.$$

Thus for every $t > 0$

$$\mu_t(B) = 1.$$

We calculate the characteristic functional χ_{μ_t} of measure μ_t with respect to Hilbert space H:

$$\chi_{\mu_t}(x) = \int e^{-i\langle x,y\rangle}\,d\mu_t(y) =$$

$$= \prod_{k=1}^{\infty} \int e^{-ix_k y_k/a_k}\,dn_{a_k t}(y_k) =$$

$$= \prod_{k=1}^{\infty} e^{-\frac{1}{2}(2ta_k)(x_k/a_k)^2} = \prod_{k=1}^{\infty} e^{-\frac{1}{2}\cdot 2t x_k^2/a_k} = e^{-\frac{1}{2}\cdot 2t|x|^2}.$$

Thus the correlation operator of measure μ_t (with respect to H) equals $2tI$ and thus (i, H, B), where $i : H \hookrightarrow B$, is a Wiener space [72].

From what has been said and from Kolmogorov's theorem [49: p. 315] it follows that given the initial conditions of process X_∞ on B, it is possible to construct realizations $t \to x_t$ of this process that are continuous functions in t with values in B. This process, which we denote in the following by $\mathcal{X}$, is also a standard Wiener process on Wiener space (i, H, B) (see [72] and [49: p. 317]).

Let π_α be the restriction of mapping $\pi_{\alpha\infty}$ to B. Then in addition to properties 1)–4) we have

5) $X_\alpha = \pi_\alpha(\mathcal{X})$ $(\forall \alpha \geq 1)$.

3.5.2

Following [60] we introduce the definition

Definition 3.5.1. A measurable function u is called superharmonic with respect to a process $\mathcal{X}$ on a set V if it is locally bounded from below and for any open set $W : \overline{W} \subset V$ and any $x \in W$

1) $M_x[u(x_{T_{B\setminus W}})] \le u(x)$,

2) $\sup_{W\downarrow\{x\}} M_x[u(x_{T_{B\setminus W}})] = u(x)$,

where $T_{B\setminus W} = \inf\{t > 0 : x_t \in B\setminus W\}$.

By $\mathscr{S}^{\mathfrak{X}}$ we shall denote the sheaf of superharmonic functions of process $\mathfrak{X}$, and correspondingly, $\mathscr{H}_{\mathscr{S}^{\mathfrak{X}}} = \mathscr{S}^{\mathfrak{X}} \cap (-\mathscr{S}^{\mathfrak{X}})$ is the sheaf of harmonic functions. Relation (5) implies the property:

(6) $\mathscr{S}_\alpha \cdot \pi_\alpha \subset \mathscr{S}^{\mathfrak{X}}$ $(\forall \alpha \ge 1)$.

Let $\mathscr{P} = (\mathscr{P}_t)_{t>0}$ be the Markov semigroup of process $\mathfrak{X}$ and let $\mathfrak{R} = (\mathfrak{R}_\lambda)_{\lambda \ge 0}$ be its resolvent. For any function $f \in \mathbb{B}^+(B)$ with bounded support $\mathrm{supp}\, f$, the function $\mathfrak{R}_0 f$ is superharmonic on B and harmonic on the set $B\setminus \mathrm{supp}\, f$. In addition, for any open set $V \subset B$, the function

$$\varphi(x) = M_x(f(x_{T_{B\setminus V}}))$$

is harmonic on V.

As Goodman showed [58], in contrast to the finite-dimensional case, a harmonic function is not necessarily continuous. Following [60] we introduce the definition

Definition 3.5.2. The function f defined on open set V is called H-continuous at point $x \in V$, if the function $F(h) = f(x + h)$ $(h \in H)$ defined on a neighborhood of zero is continuous at point $h = 0$.

According to [72: Theorem 6.2, p. 119], for any function $f \in \mathbb{B}(B)$ the function $\mathscr{P}_t f$ is H-continuous for any $t > 0$. From this analogy of the strong Feller property, as also in the finite-dimensional case, we deduce the truth of the following proposition (see [60]).

Proposition 3.5.1. *Every harmonic function on an open set V is H-continuous at all points of V.*

Definition 3.5.3. A boundary point a of an open set V is called regular if $P^a(T_{B\setminus V} = 0) = 1$. The set V is called regular if all its limit points are regular.

Any cylindrical set $V = \pi_\alpha^{-1}(V_\alpha)$ with regular base $V_\alpha \subset \mathbb{R}^\alpha$ (with respect to sheaf $\mathscr{S}_\alpha$) will also be regular in the sense of Definition 3.5.3. As in the finite-dimensional case, the Poincaré regularity criterion holds: a boundary point is regular if it can be tangent from the outside to a cone (see [72]). In particular from this criterion it follows that a ball is a regular set.

Proposition 3.5.2. *If a is a regular boundary point of an open set V, and a "boundary" function f is H-continuous at point a, then*

$$H \cdot \lim_{x \to a} M_x[f(x_{T_{B\setminus V}})] = f(a).$$

Remark 3.5.1. Let $\mathcal{X}^\lambda$, $\lambda > 0$, be the subprocess of $\mathcal{X}$ defined by semigroup $\mathcal{P}^\lambda = (e^{-\lambda t}\mathcal{P}_t)_{t>0}$. The definitions of superharmonic (harmonic) function, regular point, and the properties contained in Propositions 3.5.1 and 3.5.2 transfer word for word to the case of process $\mathcal{X}^\lambda$.

3.5.3

Let $\mathbb{Z}^\infty$ be the subgroup of $\mathbb{R}^\infty$ consisting of vectors with integer coordinates. We have

$$\mathbb{T}^\infty = \mathbb{R}^\infty / 2\pi\mathbb{Z}^\infty.$$

Let $\gamma : \mathbb{R}^\infty \to \mathbb{T}^\infty$ be the natural homomorphism, and let γ_B be the restriction of γ to B; it is obvious that γ_B is a continuous mapping of $(B, \|\cdot\|)$ to $\mathbb{T}^\infty$. It is easy to see that the transition function of process $\mathcal{X}$ is compatible in the sense of Theorem 3.3.3 with the mapping γ_B, therefore the Markov process $X = \gamma_B(\mathcal{X})$ is correctly defined.

Example 3.5.1. We recall that in Sec. 3.3 (Example 3.3.3) we considered the projective sequence of harmonic spaces $(E^\alpha, \mathcal{S}_\alpha)$:

$$E^\alpha = \mathbb{T}^\alpha, \ \ \mathcal{S}_\alpha = \mathcal{S}_{\mathcal{L}_\alpha}, \ \ \mathcal{L}_\alpha = \sum_{k=1}^{\alpha} a_k \partial_k^2.$$

Based on Example 3.3.8, it can be easily shown that limit sheaf $\mathcal{S} = \lim \mathrm{proj}\, \mathcal{S}_\alpha$ coincides with sheaf $\mathcal{S}_c^X$ of lower-semicontinuous superharmonic functions of process X. From the fact that $X = \gamma_B(\mathcal{X})$, it follows that $\mathcal{S} \cdot \gamma_B \subset \mathcal{S}^X \cdot \gamma_B \subset \mathcal{S}^{\mathcal{X}}$. Based on this inclusion we shall show that if $\sum_{k=1}^\infty 1/a_k < \infty$, then $(\mathbb{T}^\infty, \mathcal{S})$ is a harmonic space. According to Theorem 3.3.2, it is sufficient to show that the Bauer convergence property is satisfied.

Let V be an open subset of $\mathbb{T}^\infty$ and let $\{u_n\}_1^\infty$ be a monotonically increasing sequence of functions of $\mathcal{H}_{\mathcal{S}}(V)$ such that the function $u := \sup_n u_n$ is locally bounded. We shall show that the function u is continuous; consequently $u \in \mathcal{H}_{\mathcal{S}}(V)$; thus the proof will be finished. We set $\underline{V} = \gamma_B^{-1}(V)$, $\underline{u}_n = u_n \cdot \gamma_B$, and $\underline{u} = u \cdot \gamma_B$. Then $\underline{u}_n \in \mathcal{H}_{\mathcal{S}^{\mathcal{X}}}(\underline{V})$ and $\underline{u} =\uparrow \lim_{n\to\infty} \underline{u}_n \in \mathcal{H}_{\mathcal{S}^{\mathcal{X}}}(\underline{V})$. According to Proposition 3.5.1, the function $\underline{u}$ is H-continuous on $\underline{V}$. For each $x \in T^\infty$, we denote by $\underline{x}$ the unique point of set $]-\pi, \pi]^\infty \subset B$ such that $\gamma_B(\underline{x}) = x$. We have $u(x) = \underline{u}(\underline{x})$. Now the continuity of u follows from the H-continuity of $\underline{u}$ and from the fact that $[-\pi, \pi]^\infty$ is a compact subset of H. $\qquad\square$

Chapter 4
Markov processes and harmonic structures on a group

In this chapter we develop the results of Chapter 3 in the case when topological space E is a locally compact connected and locally connected abelian group with countable base. According to the well-known Dixmier theorem [63: Theorem B, p. 19], group E is isomorphic to the group $\mathbb{R}^p \times \mathbb{T}^m (m \leq \infty)$. The case $m < \infty$ does not differ much from the classical $E = \mathbb{R}^{p+m}$; therefore our principal attention will be given to the case $E = \mathbb{R}^p \times \mathbb{T}^\infty$. In Sec. 4.1, starting with a hyperharmonic translation-invariant sheaf $\mathcal{S}$, we give the infinite-dimensional analogue of Bony's theorem (see Sec. 2.6). More precisely, we shall show that there exists an infinite-dimensional elliptic differential operator $\mathcal{L}$ such that sheaf $\mathcal{H}_{\mathcal{S}}$ coincides with sheaf $\mathcal{H}_{\mathcal{L}}$ of continuous functions which satisfy the equation $\mathcal{L}u = 0$ in the sense of distribution theory. In a certain sense, differential operator $\mathcal{L}$ generates a space-homogeneous Markov process X on E with almost surely continuous trajectories. We shall show that the sheaf $\mathcal{S}^X$ of superharmonic functions of this process coincides with the original sheaf $\mathcal{S}$. Thus, in this way ($\mathcal{S} \to \mathcal{L} \to X$) we shall sharpen the well-known Meyer theorem "hyperharmonic sheaf $\to$ Markov process" in the following manner: "hyperharmonic translation-invariant sheaf $\to$ space-homogeneous Markov process." We note that in one particular case (the sheaf $\mathcal{S}$ symmetric and $(E, \mathcal{H}_{\mathcal{S}})$ a $\mathcal{P}$-Brelot space) a similar result was obtained by Forst [55], who used a different approach. In Sec. 4.2, continuing the chain $\mathcal{S} \to \mathcal{L} \to X$, we shall consider a space-homogeneous Markov process X on E with almost surely continuous trajectories, and we give conditions which guarantee the existence in the sheaf $\mathcal{S} := \mathcal{S}^X$ of the given axiomatic properties (see Secs. 2.4-2.5). With each such Markov process it is possible to associate in a natural manner a semigroup $(\mu_t)_{t>0}$ of Gaussian measures on E. The above-mentioned conditions are expressed in terms of this semigroup (absolute continuity with respect to Haar measure, existence of continuous density, etc.). Gaussian semigroups on E have very simple structure. In fact, if $(\mu_t)_{t>0}$ is such a semigroup, then for every $t > 0$ the Fourier transform $\widehat{\mu}_t$ of measure μ_t has the form

$$\widehat{\mu}_t(\theta) = \exp\{-t[\Psi(\theta) + i\ell(\theta)]\}, \ \theta \in \widehat{E},$$

where $\widehat{E} = \mathbb{R}^p \times \mathbb{Z}^{(\infty)}$ is the dual group, and ℓ and Ψ are the linear and the non-negative definite quadratic forms on $\widehat{E}$. In the natural basis of group $\widehat{E}$ these forms can be written in the form

$$\ell(\theta) = \sum_{k=1}^{\infty} b_k \theta_k,$$

$$\Psi(\theta) = \sum_{i,j=1}^{\infty} a_{ij}\theta_i\theta_j, \quad \theta = (\theta_i)_1^{\infty} \in \widehat{E}.$$

In the remaining two sections of this chapter, in essence we shall give conditions on a vector $b = (b_i)_1^{\infty}$ and a matrix $A = (a_{ij})_1^{\infty}$ which guarantee the existence of the given properties of semigroup $(\mu_t)_{t>0}$, and at the same time the existence of the given properties of sheaf $\mathcal{S}$.

As already noted, every function $u \in \mathcal{H}_{\mathcal{S}}(V)$ is a weak solution of the equation $\mathcal{L}u = 0$, where

$$\mathcal{L} = \sum_{i,j=1}^{\infty} a_{ij}\partial_i\partial_j + \sum_{i=1}^{\infty} b_i\partial_i$$

is the differential operator associated with process X. The set of weak solutions of this equation, generally speaking, is wider than the set of functions that are $\mathcal{H}_{\mathcal{S}}(V)$-harmonic on V. In the next chapter it will be shown that these sets coincide if and only if the process X is in class $\mathcal{B}$ (see Sec. 3.2). In turn, this condition can be expressed directly in terms of the coefficients of $A = (a_{ij})_1^{\infty}$ and $b = (b_i)_1^{\infty}$ of operator $\mathcal{L}$ (see Secs. 4.3 and 4.4).

4.1 Harmonic groups

4.1.1

Here and in the following we shall consider a locally compact group $E = \mathbb{R}^p \times \mathbb{T}^{\infty}$, where $\mathbb{R}^p$ is the p-dimensional Euclidean space, $\mathbb{T} = \mathbb{R}^1/2\pi\mathbb{Z}$ ($\mathbb{Z}$ is the additive group of integers) is the torus, and $\mathbb{T}^{\infty}$ is the direct product of a countable number of tori $\mathbb{T}$. We shall consider functions on E as functions on $\mathbb{R}^{\infty}$ that are invariant with respect to the actions of group $\{0\} \times \mathbb{Z}^{\infty}$ ($\{0\} \subset \mathbb{R}^p$). We set $E_n = \mathbb{R}^p \times \mathbb{T}^n$ and let π_{nm} ($m \geq n$) be the natural homomorphism of group E_m onto group E_n. The sequence of groups $\{E_n\}$ is projective with respect to system of mappings $\{\pi_{nm}\}$ ($m \geq n$), and group E is its projective limit. We shall call the set $V = \pi_{n\infty}^{-1}(V_n)$ cylindrical, and V_n is its base. We shall also assign the term "cylindrical" to functions of the type $f \cdot \pi_{n\infty}$, where f is a function on E_n (see Sec. 3.3, 3.3.2). By dx we denote the Haar measure on group E. We have:

$$dx = \otimes_{k=1}^{\infty} dx_k,$$

where dx_k $(k \le p)$ is the Lebesgue measure on $\mathbb{R}^1$ and dx_k $(k > p)$ is the Haar measure on $\mathbb{T}$ coinciding with the normed Lebesgue measure on the segment $[-\pi, \pi]$ under the identification $\mathbb{T} \approx [-\pi, \pi]$.

Let $\widehat{E}$ be the character group of E. The group $\widehat{E}$ can be identified in a natural manner with the group $\mathbb{R}^p \times \mathbb{Z}^{(\infty)}$, where $\mathbb{Z}^{(\infty)}$ is the subgroup of $\mathbb{Z}^\infty$ consisting of sequences $\theta = (\theta_k)_{k=1}^\infty$, which are eventually zero. We denote the Fourier transform of measure μ and function f on E by $\widehat{\mu}(\theta)$ and $\widehat{f}(\theta)$ $(\theta \in \widehat{E})$.

4.1.2

Let $\mathscr{S}$ be a sheaf of functions on E. We shall say that $\mathscr{S}$ is translation-invariant if for any open set $V \subset E$, function $u \in \mathscr{S}(V)$, and $a \in E$, the function $f_a(x) = f(x+a)$ defined on the set $V - a$ belongs to the set $\mathscr{S}(V - a)$.

Let $\mathscr{S}$ be a hyperharmonic sheaf on E. We shall call $(E, \mathscr{S})$ a *harmonic group* if the following conditions are satisfied:
1) $(E, \mathscr{S})$ is a harmonic space,
2) $\mathscr{S}$ is translation-invariant.

Analogously, if $\mathscr{H}$ is a harmonic sheaf on E, then we shall call $(E, \mathscr{H})$ a *Bauer* (respectively, *Brelot*) *harmonic group* if:
1) $(E, \mathscr{H})$ is a harmonic Bauer (Brelot) space;
2) $\mathscr{H}$ is translation-invariant.

We shall call the harmonic group $(E, \mathscr{S})$ an *elliptic harmonic group* if the harmonic space $(E, \mathscr{S})$ is elliptic [46: p. 66], i.e. if the following condition is satisfied: there exists a base of topology $\mathscr{F}$ consisting of resolutive sets and such that $\forall V \in \mathscr{F}, \forall x \in V, \operatorname{supp} \mu_x^V = \partial V$.

In correspondence with the general theory [46: Proposition 3.1.4], an elliptic harmonic group can be characterised as a harmonic group for which one of the following three equivalent conditions is satisfied:
1) any non-negative hyperharmonic function on a connected open set is either finite on a dense subset or is identically equal to $+\infty$;
2) there exists a base of a topology consisting of regular sets, and any non-negative harmonic function on a connected open set is either strictly positive or identically zero;
3) any non-negative hyperharmonic function on a connected open set is either strictly positive or identically zero.

We note further that if $(E, \mathscr{H})$ is a Brelot harmonic group, then setting $\mathscr{S} := \mathscr{H}^*$ (see Sec. 2.5) we get an elliptic harmonic group $(E, \mathscr{S})$, where $\mathscr{H}_{\mathscr{S}} = \mathscr{H}$. However, the converse is not true, i.e. if $(E, \mathscr{S})$ is an elliptic harmonic group, then in correspondence with property 2) and [46: Corollary 3.1.2], it is possible only to assert that $(E, \mathscr{H}_{\mathscr{S}})$ is an elliptic harmonic Bauer group, which in the general case is not a Brelot harmonic group (see [46: Remark, p. 67], and Remark 4.4.1).

4.1.3

We denote by $\mathscr{D}$ the set of cylindrical infinitely differentiable functions on E with compact supports. We shall consider the linear differential operator $\mathscr{L}$ on $\mathscr{D}$ with constant coefficients

$$\mathscr{L} = \sum_{i,j=1}^{\infty} a_{ij}\partial_i\partial_j + \sum_{i=1}^{\infty} b_i\partial_i.$$

It is obvious that $\mathscr{L} : \mathscr{D} \to \mathscr{D}$. We shall call the function $u \in L_{1\,\mathrm{loc}}(E, dx)$ a weak solution of equation $\mathscr{L}u = 0$ on V if the condition

$$\langle u, \mathscr{L}^*\varphi \rangle = 0, \;\; \forall \varphi \in \mathscr{D} : \mathrm{supp}\,\varphi \subset V,$$

is satisfied, where $\langle \cdot, \cdot \rangle$ is the scalar product on $L_2(E, dx)$, and $\mathscr{L}^*$ is the differential operator formally conjugate to $\mathscr{L}$. We denote by $\mathscr{H}_{\mathscr{L}}(V)$ the set of functions $u \in \mathbb{C}(V)$ which are weak solutions of equation $\mathscr{L}u = 0$ in V. It is obvious that $\mathscr{H}_{\mathscr{L}}$ is a translation-invariant harmonic sheaf on E.

We shall call the operator $\mathscr{L}$ elliptic if the following conditions are satisfied:
1) $a_{ij} = a_{ji}, \forall i, j = 1, 2, \ldots,$
2) $(a_{ij})_1^n > 0, \forall n = 1, 2, \ldots$.

Theorem 4.1.1. *Let $(E, \mathscr{S})$ be an elliptic harmonic group such that $1 \in \mathscr{H}_{\mathscr{S}}(E)$. Then there exists an elliptic differential operator $\mathscr{L}$ that is unique up to a constant factor, such that $\mathscr{H}_{\mathscr{S}} = \mathscr{H}_{\mathscr{L}}$.*

4.1.4

Theorem 4.1.1 can be considered as the infinite-dimensional analogue of Bony's theorem for translation-invariant sheaves in Euclidean space $\mathbb{R}^p$ (see Sec. 2.6). We shall state Bony's theorem in the following convenient form:

Bony's Theorem. *Let $(\mathbb{R}^p, \mathscr{S})$ be an elliptic harmonic group such that $1 \in \mathscr{H}_{\mathscr{S}}(\mathbb{R}^p)$. Then there exists an elliptic differential operator $\mathscr{L}$ with constant coefficients, unique up to a constant factor, such that $\mathscr{H}_{\mathscr{S}} = \mathscr{H}_{\mathscr{L}}$. In addition:*
1) $\mathscr{H}_{\mathscr{L}} \subset \mathbb{C}^\infty,$
2) $(\mathbb{R}^p, \mathscr{H}_{\mathscr{S}})$ *is a Brelot harmonic group.*

Thus, in the finite-dimensional case the concept of elliptic harmonic group and Brelot harmonic group in essence coincide. In the infinite-dimensional case, as noted above, this is not so (see Theorem 4.4.3).

We consider the group $\mathbb{R}^p \times \mathbb{T}^m = \mathbb{R}^{p+m}/\{0\} \times 2\pi\mathbb{Z}^m$ and let γ be the natural homomorphism of group $\mathbb{R}^{p+m}$ onto group $\mathbb{R}^p \times \mathbb{T}^m$. The mapping γ has the following property: for any $x \in \mathbb{R}^{p+m}$ and $y \in \mathbb{R}^p \times \mathbb{T}^m$ such that $\gamma(x) = y$, there exist neighborhoods V of point x and W of point y such that $\gamma(V) = W$, and the restriction of γ to V is a diffeomorphism of V into W. This property permits us to establish a one-to-one correspondence between translation-invariant sheaves on groups $\mathbb{R}^{p+m}$ and $\mathbb{R}^p \times \mathbb{T}^m$. In fact, let $\mathcal{S}$ be a translation-invariant sheaf on group $\mathbb{R}^p \times \mathbb{T}^m$. Let $B_\varepsilon(0)$ be an open ball of radius $\varepsilon < 1$ with center at zero; it is obvious that $\gamma : B_\varepsilon(0) \to \gamma(B_\varepsilon(0))$ is a diffeomorphism. Now let V be an open subset of $\mathbb{R}^{p+m}$; we denote by $\widetilde{\mathcal{S}}(V)$ the set of functions u on V having the following property: for any $x \in V$ and any ball $B_\varepsilon(x) \subset V$, the function $u_x(y) = u(y + x)$ defined on $B_\varepsilon(0)$ belongs to the set $\mathcal{S}(\gamma[B_\varepsilon(0)]) \cdot \gamma$. It is obvious that $\widetilde{\mathcal{S}}$ is a translation-invariant sheaf on the group $\mathbb{R}^{p+m}$. It is also clear that if $(\mathbb{R}^p \times \mathbb{T}^m, \mathcal{S})$ is an elliptic harmonic group, then $(\mathbb{R}^{p+m}, \widetilde{\mathcal{S}})$ is also an elliptic harmonic group. Now applying Bony's theorem, we get

Theorem 4.1.2. *Let* $(\mathbb{R}^p \times \mathbb{T}^m, \mathcal{S})$ *be an elliptic harmonic group such that* $1 \in \mathcal{H}_{\mathcal{S}}(\mathbb{R}^p \times \mathbb{T}^m)$. *Then there exists an elliptic differential operator* $\mathcal{L}$ *with constant coefficients unique up to a constant factor such that* $\mathcal{H}_{\mathcal{S}} = \mathcal{H}_{\mathcal{L}}$. *In addition:*
1) $\mathcal{H}_{\mathcal{L}} \subset \mathbb{C}^\infty$,
2) $(\mathbb{R}^p \times \mathbb{T}^m, \mathcal{H}_{\mathcal{S}})$ *is a Brelot harmonic group.*

4.1.5

Now we go to the proof of Theorem 4.1.1. We fix m and we define the hyperharmonic sheaf $\mathcal{S}_m$ on group $E_m = \mathbb{R}^p \times \mathbb{T}^m$ in the following manner: for every open set $V \subset E_m$ we set

$$\mathcal{S}_m(V) := \{u : u \cdot \pi_{m\infty} \in \mathcal{S}(\pi_{m\infty}^{-1}(V))\}.$$

It is obvious that we thus in fact get a hyperharmonic sheaf on the group E_m which is also translation-invariant. We shall show that $(E_m, \mathcal{S}_m)$ is an elliptic harmonic group, and then applying Theorem 4.1.2 (it is obvious that $1 \in \mathcal{H}_{\mathcal{S}_m}(E_m)$), we get the elliptic differential operator

$$\mathcal{L}_m = \sum_{i,j=1}^{m} a_{ij}^m \partial_i \partial_j + \sum_{j=1}^{m} b_j^m \partial_j$$

such that $\mathcal{H}_{\mathcal{S}_m} = \mathcal{H}_{\mathcal{L}_m}$. From the relation $\mathcal{S}_m \cdot \pi_{mn} \subset \mathcal{S}_n$ $(m < n)$ it follows that a sequence of differential operators $\{\mathcal{L}_n\}_1^\infty$ can be selected in such a way that the following condition is satisfied:

$$\mathcal{L}_n(\varphi \cdot \pi_{mn}) = (\mathcal{L}_m \varphi) \cdot \pi_{mn}.$$

In particular, we therefore get:

$$a_{ij}^m = a_{ij}^n,$$
$$b_i^m = b_i^n, \quad \forall i, j \le m, \ \forall m, n : m \le n.$$

We set:

$$a_{ij} := a_{ij}^n,$$
$$b_i := b_i^n, \quad n \geq \max(i, j),$$

and consider the infinite-dimensional differential operator

$$\mathcal{L} = \sum_{i,j=1}^{\infty} a_{ij} \partial_i \partial_j + \sum_{i=1}^{\infty} b_i \partial_i.$$

It is obvious that $\mathcal{L}$ is an elliptic differential operator (see Sec. 4.1.3), and in addition

$$\mathcal{L}(\varphi \cdot \pi_{n\infty}) = (\mathcal{L}_n \varphi) \cdot \pi_{n\infty} \ (\forall n \geq 1, \ \forall \varphi \in \mathbb{C}_k^{\infty}(E_n)).$$

We shall show that $\mathcal{H}_{\mathcal{L}} = \mathcal{H}_{\mathcal{S}}$. Let $\tilde{\nu}_n$ be the Haar measure of closed subgroup $\{0\} \times \mathbb{T}^{\infty}$, $\{0\} \subset \mathbb{R}^p \times \mathbb{T}^n$ and $\nu_n := \varepsilon_0 \otimes \tilde{\nu}_n$ a measure on E; we have $\nu_n \to \varepsilon_0$, $n \to \infty$. Let V be an open cylindrical subset of E such that $V = \pi_{n\infty}^{-1}(V_n)$ and $u \in \mathcal{H}_{\mathcal{L}}(V)$; we set $u_m(x) = u * \nu_m(x)$, $m > n$, $x \in V$. For every m the function u_m is cylindrical: $u_m = \bar{\bar{u}}_m \cdot \pi_{m\infty}$, and in addition for any function $\varphi \in \mathbb{C}_k^{\infty}(V_m)$, $V_m = \pi_{m\infty}(V)$:

$$\langle \bar{\bar{u}}_m, \mathcal{L}_m^* \varphi \rangle = \langle u_m, \mathcal{L}^*(\varphi \cdot \pi_{m\infty}) \rangle =$$
$$= \langle u, \mathcal{L}^*(\varphi \cdot \pi_{m\infty}) \rangle = 0.$$

Consequently $\bar{\bar{u}}_m \in \mathcal{H}_{\mathcal{L}_m}(V_m) = \mathcal{H}_{\mathcal{S}_m}(V_m)$, and thus $u_m \in \mathcal{H}_{\mathcal{S}}(V)$. Now it remains to note that the sequence u_m converges to u locally uniformly on the set V. Thus $u \in \mathcal{H}_{\mathcal{S}}(V)$ and thus the inclusion $\mathcal{H}_{\mathcal{L}} \subset \mathcal{H}_{\mathcal{S}}$ is proved. Let $u \in \mathcal{H}_{\mathcal{S}}(V)$. Since the sheaf $\mathcal{H}_{\mathcal{S}}$ is translation-invariant, then for every $m \geq 1$ we have $u_m \in \mathcal{H}_{\mathcal{S}}(V)$, and consequently $\bar{\bar{u}}_m \in \mathcal{H}_{\mathcal{S}_m}(V_m)$. Now, arguing as in the previous case, we arrive at the reverse inclusion $\mathcal{H}_{\mathcal{S}} \subset \mathcal{H}_{\mathcal{L}}$.

Thus, the proof of Theorem 4.1.1 reduces to a proof of the fact that for every m, $(E_m, \mathcal{S}_m)$ is an elliptic harmonic group. According to Sec. 4.1.2, this fact is equivalent to the fact that $(E_m, \mathcal{H}_{\mathcal{S}_m})$ is an elliptic harmonic Bauer group. We note that the axiom of nondegeneracy and the axiom of Bauer convergence for $\mathcal{H}_{\mathcal{S}_m}$ are satisfied trivially, and therefore it remains to verify the axiom of existence of a base of regular sets (see Sec. 2.5). Since $\mathcal{H}_{\mathcal{S}_m}$ is translation-invariant, it is sufficient to establish the existence of a base of neighborhoods of the neutral element consisting of regular sets. We shall present the proof of the existence of such a base in several steps.

1) We shall consider the open cylindrical set $V = \pi_{m\infty}^{-1}(V_m)$ and let ${}^V R_f$ be the reduction of function f with respect to harmonic space $(V, \mathcal{S})$, i.e.

$$^V R_f = \inf\{u \in \mathcal{S}^+(V) : u \geq f\}.$$

We shall show that for any function φ on V_m, we have ${}^V R_{\varphi \cdot \pi_{m\infty}} = g \cdot \pi_{m\infty}$, where g is a function on V_m. We shall denote by $u \cdot \sigma$ the translation of function u by an element σ of group $\{0\} \times \mathbb{T}^\infty$, $\{0\} \subset \mathbb{R}^p \times \mathbb{T}^m$. If $u \in \mathcal{S}^+(V)$ such that $u \geq \varphi \cdot \pi_{m\infty}$, then $u \cdot \sigma \geq \varphi \cdot \pi_{m\infty} \cdot \sigma = \varphi \cdot \pi_{m\infty}$, and therefore $u \cdot \sigma \geq {}^V R_{\varphi \cdot \pi_{m\infty}}$. Therefore ${}^V R_{\varphi \cdot \pi_{m\infty}} \cdot \sigma \geq {}^V R_{\varphi \cdot \pi_{m\infty}}$. Analogously, ${}^V R_{\varphi \cdot \pi_{m\infty}} \cdot \sigma^{-1} \geq {}^V R_{\varphi \cdot \pi_{m\infty}}$, and therefore ${}^V R_{\varphi \cdot \pi_{m\infty}} \geq {}^V R_{\varphi \cdot \pi_{m\infty}} \cdot \sigma$. Consequently $\forall \sigma \in \{0\} \times \mathbb{T}^\infty$, we have ${}^V R_{\varphi \cdot \pi_{m\infty}} \cdot \sigma = {}^V R_{\varphi \cdot \pi_{m\infty}}$. Therefore there exists a function g on V_m such that ${}^V R_{\varphi \cdot \pi_{m\infty}} = g \cdot \pi_{m\infty}$.

2) Now let $V = \pi_{m\infty}^{-1}(V_m)$ be a relatively compact open neighborhood of zero, such that $\partial V = \partial \overline{V}$. We choose a function $\varphi \in \mathbb{C}_k^+(V_m)$ such that $\varphi \leq 1$ and $\varphi(0) = 1$. Then according to [46: Proposition 2.2.3] the function ${}^V R_{\varphi \cdot \pi_{m\infty}}$ is continuous, superharmonic, does not exceed 1, equals 1 on the set $\pi_{m\infty}^{-1}(0)$ and is harmonic on the set $V \backslash \operatorname{supp}(\varphi \cdot \pi_{m\infty}) = \pi_{m\infty}^{-1}(V_m \backslash \operatorname{supp} \varphi)$. We define the function u on the set $E \backslash \operatorname{supp}(\varphi \cdot \pi_{m\infty})$:

$$
u(x) = \begin{cases}
{}^V R_{\varphi \cdot \pi_{m\infty}}(x), & x \in V \backslash \operatorname{supp}(\varphi \cdot \pi_{m\infty}), \\
0, & x \in E \backslash \overline{V}, \\
\limsup\limits_{z \to x,\, z \in V} {}^V R_{\varphi \cdot \pi_{m\infty}}(z), & x \in \partial V
\end{cases}
$$

and we shall show that it is subharmonic on this set. We shall argue as in [46: p. 67]. It is obvious that the function u is upper-semicontinuous. Let W be a relatively compact resolutive set such that $\overline{W} \subset E \backslash \operatorname{supp}(\varphi \cdot \pi_{m\infty})$ and let $v \in \mathcal{S}_u$ (see Sec. 2.3). We define a function u^* on V:

$$
u^*(x) = \begin{cases}
{}^V R_{\varphi \cdot \pi_{m\infty}}(x), & x \in V \backslash W \\
\inf({}^V R_{\varphi \cdot \pi_{m\infty}}, v)(x), & x \in V \cap W.
\end{cases}
$$

The function u^* is lower-semicontinuous on V, therefore by [46: Proposition 2.1.2], it is superharmonic on V, and in addition $u^* \geq \varphi \cdot \pi_{m\infty}$. Therefore $u^* \geq {}^V R_{\varphi \cdot \pi_{m\infty}}$ and in particular

$$
{}^V R_{\varphi \cdot \pi_{m\infty}}(x) \leq v(x), \quad \forall x \in V \cap W.
$$

Consequently, $\forall x \in V \cap W$ we have

$$
{}^V R_{\varphi \cdot \pi_{m\infty}}(x) \leq \inf \{v(x) : v \in \mathcal{S}_u\} = \overline{H}_u^W(x).
$$

According to [46: Proposition 2.4.1], the function $\overline{H}_u^W$ is harmonic on W, and therefore

$$
u(x) \leq \overline{H}_u^W(x), \quad \forall x \in W.
$$

According to [46: Theorem 1.2.1],

$$
\mu_x^W(u) \leq \overline{H}_u^W(x), \quad \forall x \in W;
$$

on the other hand, if g is a continuous function such that $u \leq g$, then by [46: p. 18] we have

$$\overline{H}_u^W \leq \overline{H}_g^W = H_g^W = \mu.^W(g).$$

Passing to infimum on the right-hand side of this inequality over all g and considering the upper-semicontinuity of u, we get

$$\mu_x^W(u) \geq \overline{H}_u^W(x), \quad \forall x \in W.$$

Thus finally we get

$$u(x) \leq \overline{H}_u^W(x) = \mu_x^W(u), \quad \forall x \in W.$$

In view of the arbitrariness of the choice of resolutive set $W \subset E \backslash \operatorname{supp}(\varphi \cdot \pi_{m\infty})$, the proof that u is subharmonic is complete.

3) We shall consider the set

$$A = \{x \in V : {}^V R_{\varphi \cdot \pi_{m\infty}}(x) = 1\}.$$

By 1), A is a cylindrical set: $A = \pi_{m\infty}^{-1}(A_m)$, and in addition $\pi_{m\infty}^{-1}(0) \subset A$. We shall show that A is a compact subset of V. We assume the contrary, i.e. $\overline{A} \cap \partial V \neq \emptyset$ and let x be a point of this set. If W is a connected open neighborhood of x such that $W \cap \operatorname{supp}(\varphi \cdot \pi_{m\infty}) = \emptyset$, then the function $1 - u$ is non-negative, superharmonic on W, and approaches zero at x. By the ellipticity property (see Sec. 4.1.2) this function is identically zero on W — a contradiction, since $W \backslash \overline{V} \neq \emptyset$ and $(1 - u)|_{W \backslash \overline{V}} \equiv 1$.

4) We choose an open cylindrical neighborhood $W = \pi_{m\infty}^{-1}(W_m)$ of set A such that $\overline{W} \subset V$. Since

$$^V R_{\varphi \cdot \pi_{m\infty}}(x) < 1, \quad \forall x \in \partial W,$$

then

$$\sup_{\partial W} {}^V R_{\varphi \cdot \pi_{m\infty}} := \varepsilon < 1.$$

For any number $\alpha : \varepsilon < \alpha < 1$, we consider the set

$$^\alpha V := \{x \in W : {}^V R_{\varphi \cdot \pi_{m\infty}}(x) > \alpha\}.$$

It is clear that $^\alpha V$ is an open cylindrical neighborhood of $A : {}^\alpha V = \pi_{m\infty}^{-1}({}^\alpha V_m)$, where $^\alpha \overline{V} \subset W$. In addition, $^\alpha V$ is regular, since the function

$$u_\alpha := {}^V R_{\varphi \cdot \pi_{m\infty}} - \alpha$$

is a barrier at all boundary points of $^\alpha V$ at once (see Sec. 2.5). Now we note that the regularity of cylindrical set $^\alpha V$ with respect to $\mathcal{H}_\varphi$ is equivalent to the regularity of its base $^\alpha V_m$ with respect to $\mathcal{H}_{\varphi_m}$. And finally, let $V \downarrow \pi_{m\infty}^{-1}(0)$; then it is obvious that $^\alpha V_m \downarrow \{0\}$, and consequently $\{^\alpha V_m\}$ is the desired base of regular neighborhoods of zero. $\qquad \square$

4.1.6

We shall call a Markov process X a space-homogeneous Markov process if its transition function $p(t; x, \Gamma)$ is invariant with respect to the action of a group E:

$$p(t; x + y, \Gamma + y) = p(t; x, \Gamma) \ (\forall x, y \in E, \ \Gamma \in \mathcal{B}(E)).$$

Theorem 4.1.3. *Let $(E, \mathcal{S})$ be an elliptic harmonic group such that $1 \in \mathcal{H}_{\mathcal{S}}(E)$. Then there exists a space-homogeneous Markov process X with almost surely continuous trajectories such that sheaf $\mathcal{S}^X$ of its superharmonic functions coincides with $\mathcal{S}$.*

Proof. Let

$$\mathcal{L} = \sum_{i,j=1}^{\infty} a_{ij} \partial_i \partial_j + \sum_{i=1}^{\infty} b_i \partial_i$$

be the elliptic differential operator associated by Theorem 4.1.1 with elliptic harmonic group $(E, \mathcal{S})$. On the dual group $\widehat{E} = \mathbb{R}^p \times \mathbb{Z}^{(\infty)}$ we define the functions

$$\Psi(\theta) = \sum_{i,j=1}^{\infty} a_{ij} \theta_i \theta_j,$$

$$\ell(\theta) = \sum_{i=1}^{\infty} b_i \theta_i, \ \ \theta = (\theta_i)_1^{\infty} \in \widehat{E}$$

and for every $t > 0$ we consider the function

$$\theta \to \exp\{-t[\Psi(\theta) - i\ell(\theta)]\}.$$

According to [29: 7.19, 7.20, 8.4] this function is positive definite, and therefore by Bochner's theorem there exists a probability measure μ_t on E such that

$$\widehat{\mu}_t(\theta) = \exp\{-t[\Psi(\theta) - i\ell(\theta)]\}.$$

The family of measures $(\mu_t)_{t>0}$ has the semigroup convolution property $\mu_t * \mu_s = \mu_{t+s}$, and for $t \downarrow 0$, weak convergence of measures $\mu_t \to \varepsilon_0$ takes place. We set

$$p(t; x, \Gamma) = \mu_t(x - \Gamma) \ (t > 0, \ x \in E, \ \Gamma \in \mathcal{B}(E));$$

it is obvious that $p(t; x, \Gamma)$ is a transition function. It satisfies the conditions of Theorem 3.14 of [54] and therefore generates a standard Markov process X on phase space $(E, \mathcal{B}(E))$. By construction, X is a space-homogeneous process. By [29, 18.27], the following condition is satisfied:

$$\lim_{t \to 0} \frac{1}{t} \mu_t(E \backslash V) = 0 \text{ for all open neighborhoods } V \text{ of } 0,$$

and consequently by Theorem 3.5 of [54] the process X has almost surely continuous trajectories. We note finally that on a set $\mathcal{D}$ of cylindrical infinitely differentiable functions with compact supports, the infinitesimal operator of process X coincides with differential operator $\mathscr{L}$.

Let $\mathscr{S}^X$ be the sheaf of superharmonic functions of process X; we shall show that $\mathscr{S}^X = \mathscr{S}$. For every m we consider a Markov process $X_m = \pi_{m\infty}(X)$ on group $E_m = \mathbb{R}^p \times \mathbb{T}^m$ (see Sec. 3.3.4). On the functions $\varphi \in \mathbb{C}_k^\infty(E_m)$ the infinitesimal operator of X_m coincides with the differential operator

$$\mathscr{L}_m = \sum_{i,j=1}^m a_{ij}\partial_i\partial_j + \sum_{i=1}^m b_i\partial_i.$$

Now applying Theorem 13.9 of [54] we conclude that $\mathscr{H}_{\mathscr{S}^{X_m}} = \mathscr{H}_{\mathscr{L}_m}$. On the other hand, $\mathscr{H}_{\mathscr{L}_m} = \mathscr{H}_{\mathscr{S}_m}$ (see Sec. 4.1.5); consequently $\mathscr{H}_{\mathscr{S}^{X_m}} = \mathscr{H}_{\mathscr{S}_m}$. Therefore, in turn, we conclude that $\mathscr{S}^{X_m} = \mathscr{S}_m$. The sequence of processes $\{X_m\}_1^\infty$ is projective and process X is its projective limit. According to Proposition 1.3.2, we have

$$\mathscr{S} = \lim\mathrm{proj}\,\mathscr{S}_m = \mathscr{S}_c^X,$$

where $\mathscr{S}_c^X$ is the sheaf of lower-semicontinuous superharmonic functions of X. According to Remark 3.3.3, $\mathscr{S}_c^X = \mathscr{S}^X$. The proof is complete. $\square$

4.2 Space-homogeneous processes and harmonic functions

In this section the starting point of our investigation is the space-homogeneous Markov process, and the ending point is the elliptic harmonic group. Thus, considering the results of the preceding section, we arrive at a scheme well-known in classical potential theory (see [53]):

| elliptic harmonic group | $\rightarrow$ | elliptic differential operator |

$$\uparrow \qquad\qquad\qquad\qquad \downarrow$$

space-homogeneous Markov process

Along with this, continuing the analogy with classical theory [53: Chapters XV-XIX], we consider in Section 4.2.8 the space-time Markov process $\dot{X}$, parabolic differential operator $\dot{\mathscr{L}} = \mathscr{L} - \partial_t$, and the "parabolic" harmonic group $(\dot{E}, \dot{\mathscr{S}})$ corresponding to these objects.

4.2.1

Let $X = (x_t, \infty, \mathcal{M}_t, P^x)$ be a space-homogeneous nonterminating Markov process on E with almost surely continuous trajectories. In correspondence with Sec. 4.1.6, this means that the transition function $p(t; x, \Gamma)$ of process X is invariant with respect to the action of the group: $p(t; x + y, \Gamma + y) = p(t; x, \Gamma)$. We set

$$\mu_t(\Gamma) = p(t; 0, -\Gamma), \quad r^\lambda(\Gamma) := \int_0^\infty e^{-\lambda t} \mu_t(\Gamma)\, dt.$$

Then, if (P_t) and (R^λ) are the Markov semigroup and the resolvent of process X, then

$$P_t f = \mu_t * f, \quad R^\lambda f = r^\lambda * f.$$

The family of probability measures $(\mu_t)_{t>0}$ has the semigroup convolution property $\mu_t * \mu_s = \mu_{t+s}$, and for $t \downarrow 0$, weak convergence of measures $\mu_t \to \varepsilon_0$ takes place, where ε_0 is the Dirac measure at zero. The properties of these semigroups of measures are listed in detail in [29] and [63], and our treatment shall be based on these sources. As a consequence of the continuity of the trajectories of process X, for every $t > 0$ the measure μ_t is Gaussian [63: Theorem 5.6.4]; we write its Fourier transform

$$\widehat{\mu}_t(\theta) = \exp\{-t[\Psi(\theta) - i\ell(\theta)]\}, \quad \theta \in \widehat{E}, \tag{4.2.1}$$

where $\Psi(\theta) \geq 0$ is a quadratic form, and $\ell(\theta)$ is a linear form on $\widehat{E}$ (see also [29: Theorem 18.17], [6-7]). Let $\Psi(\theta) = A\theta \cdot \theta$ and $\ell(\theta) = b \cdot \theta$ be representations of these forms in the natural basis of group $\widehat{E} = \mathbb{R}^p \times \mathbb{Z}^{(\infty)}$. We shall call the matrix $A = (a_{ij})_1^\infty$ the diffusion matrix, and the vector $b = (b_i)_1^\infty \in \mathbb{R}^\infty$ is the drift vector of process X.

It is not difficult to see that the Markov semigroup (P_t) of process X transforms Banach space $\mathbb{C}_0$ into itself and forms the continuous semigroup of contractions there; we denote its $\mathbb{C}_0$-generator by $\mathcal{L}_0$. From formula (4.2.1) it follows that on smooth cylindrical functions the operator $\mathcal{L}_0$ coincides with the differential operator

$$\mathcal{L} = \sum_{i,j=1}^\infty a_{ij}\partial_i\partial_j + \sum_{i=1}^\infty b_i\partial_i,$$

where ∂_i is the operator of differentiation with respect to variable $x_i : x = (x_i)_1^\infty \in E$.

4.2.2

We shall call process X transient if the following condition is satisfied: for any compact set $K \subset E$,

$$r(K) = \int_0^\infty \mu_t(K)\, dt < \infty.$$

According to the results of Port and Stone [88: Part II] (see also [29: Sec. 13.17]) the transience of X is equivalent to local integrability of the function

$\mathrm{Re}(A\theta \cdot \theta - ib \cdot \theta)^{-1}$. Thus without difficulty we get that the process X is transient if and only if one of the following conditions is satisfied:

1) $p \geq 1$, $b_1^2 + \ldots + b_p^2 \neq 0$;

2) $p \geq \mathrm{rang}\,(a_{ij})_1^p \geq 3$.

Let γ be the natural homomorphism of group $\mathbb{R}^{p+n}$ onto group $\mathbb{R}^p \times \mathbb{T}^n$ and e the identity isomorphism of $\mathbb{T}^\infty$ onto itself. We let $\tilde{\gamma} = \gamma \times e$ be the homomorphism of group $\mathbb{R}^{p+n} \times \mathbb{T}^\infty$ onto group $\mathbb{R}^p \times \mathbb{T}^n \times \mathbb{T}^\infty \simeq \mathbb{R}^p \times \mathbb{T}^\infty$. Along with process X on group $E = \mathbb{R}^p \times \mathbb{T}^\infty$, we shall consider a process X^* on group $E^* = \mathbb{R}^{p+n} \times \mathbb{T}^\infty$ with the same diffusion matrix A and drift vector b as for process X. Based on the theorem on transformation of phase space (Sec. 3.3.4), it is not difficult to verify that there exists a one-to-one correspondence between sets of space-homogeneous processes with continuous trajectories on groups E and E^*, which is given by the formula $X = \tilde{\gamma}(X^*)$. We note now that for $n \geq 3$ the process X^* is transient (if $(a_{ij})_1^{p+n} > 0$ or $(b_i)_1^{p+n} \neq 0$); we shall use this fact below.

4.2.3

The process $\widehat{X} = (\widehat{x}_t, \infty, \mathcal{M}_t, P^x)$, where $\widehat{x}_t = 2x - x_t$, is space-homogeneous, has almost surely continuous trajectories and, as it is not difficult to see, is dual to X with respect to Haar measure. Its corresponding semigroup $(\widehat{\mu}_t)$ and resolvent $(\widehat{r}^\lambda)$ are generated by the semigroup and resolvent of process X in the following manner:

$$\widehat{\mu}_t(\Gamma) = \mu_t(-\Gamma), \quad \widehat{r}^\lambda(\Gamma) = r^\lambda(-\Gamma).$$

We shall call the process X symmetric if X and $\widehat{X}$ are equivalent. The semigroup and resolvent corresponding to the symmetric process consist of symmetric measures:

$$\mu_t(-\Gamma) = \mu_t(\Gamma), \quad r^\lambda(-\Gamma) = r^\lambda(\Gamma).$$

This property, as it is not difficult to see, is equivalent to the fact that the drift vector b equals zero.

Now let X be a space-homogeneous process with diffusion matrix A and drift vector b. We shall denote by X_s the symmetric process with diffusion matrix A. We shall provide objects associated with process X_s with the index "s", for example μ_t^s, r_s^λ, etc. Formula (4.2.1) implies the relation

$$\mu_t(\Gamma) = \mu_t^s(\Gamma + b_t), \tag{4.2.2}$$

where b_t denotes the image of vector $b \cdot t \in \mathbb{R}^\infty$ for the natural homomorphism of group $\mathbb{R}^\infty$ onto group $E = \mathbb{R}^\infty/\{0\} \times \mathbb{Z}^\infty$, $\{0\} \subset \mathbb{R}^p$.

4.2.4

We introduce some conditions of absolute continuity of semigroup (μ_t) and resolvent (r^λ) with respect to Haar measure. For completeness we shall include the Cauchy semigroup (q_t) in consideration (see [29: 9.23], [32])

$$q_t = \int_0^\infty \mu_s h_t(s)\,ds, \quad h_t(s) = (4\pi)^{-1/2} t \cdot s^{-3/2} \exp\left(-\frac{t^2}{4s}\right).$$

Proposition 4.2.1. *Let the following condition be satisfied:*

$$\sup_{\theta \in \widehat{E}} |b \cdot \theta| / (A\theta \cdot \theta + 1) < \infty. \tag{C}$$

Then the following assertions are equivalent:
1) measure μ_t is absolutely continuous with respect to Haar measure for all $t > 0$;
2) measure r^λ is absolutely continuous with respect to Haar measure for all $\lambda > 0$ (in the transient case, $\lambda \geq 0$);
3) measure q_t is absolutely continuous with respect to Haar measure for all $t > 0$.
In the case of absolute continuity, the densities in 1)–3) are lower-semicontinuous and strictly positive.

Proof. 1) $\Leftrightarrow$ 2) This is the special case of the general Silverstein's theorem [98: Theorem 3.2] on simultaneous absolute continuity of the transition function and the resolvent with respect to a measure which realizes duality (i.e. a measure with respect to which the given process has a dual).
1) $\Rightarrow$ 3) Obvious.
3) $\Rightarrow$ 2): Let $(Q_t)_{t>0}$ be the semigroup of convolution operators in $\mathbb{B}$ corresponding to semigroup of measures $(q_t)_{t>0}$. From the relation

$$Q_t f = \int_0^\infty P_s f h_t(s)\,ds$$

it follows that if the function f is excessive with respect to semigroup $(P_t)_{t>0}$, then it is excessive with respect to semigroup $(Q_t)_{t>0}$. From 3) it follows that for all $t > 0$ we have $Q_t : \mathbb{B} \to \mathbb{C}_b$, and thus in turn the lower-semicontinuity of any (Q_t)-excessive function follows. In particular, any (P_t)-excessive function is lower-semicontinuous. Now let A be a Borel subset of E whose Haar measure equals zero; we shall consider the function $f = R1_A$ (for simplicity we shall assume that X is transient). We have $\forall \varphi \in \mathbb{C}_k(E)$

$$\langle R1_A, \varphi \rangle = \langle 1_A, \widehat{R}\varphi \rangle = 0,$$

consequently $f = 0$ (dx-almost everywhere). On the other hand, the function f is (P_t)-excessive and therefore lower-semicontinuous. Consequently $f \equiv 0$. In particular $r(A) = f(0) = 0$.

Now let $d\mu_t(x) = \mu_t(x)\,dx$, $\forall t > 0$; we set

$$\widetilde{\mu}_t(x) := P_s\mu_\Delta(x), \quad s + \Delta = t.$$

The function $\widetilde{\mu}_t$ does not depend on the choice of s and $\Delta : s+\Delta = t$, and coincides (dx-almost everywhere) with function $\mu_t(x)$. The fact $P_t : \mathbb{B} \to \mathbb{C}_b$ implies the lower-semicontinuity of function $\widetilde{\mu}_t$. We shall show that the function $\widetilde{\mu}_t$ is strictly positive. For each m we define a measure μ_t^m on group E_m:

$$\mu_t^m(\Gamma) := \mu_t(\pi_{m\infty}^{-1}(\Gamma)), \quad \Gamma \in \mathcal{B}(E_m).$$

We have:

$$\widehat{\mu}_t^m(\theta) = \exp\{-t(A_m\theta \cdot \theta - ib^m \cdot \theta)\}, \quad \theta \in \widehat{E}_m = \mathbb{R}^p \times \mathbb{Z}^m,$$

where $A_m = (a_{ij})_1^{m+p}$, $b^m = (b_i)_1^{m+p}$. Thus μ_t^m is a Gaussian measure on E_m. The absolute continuity of measure μ_t implies the absolute continuity of measure μ_t^m. Let γ_m be the natural homomorphism of $\mathbb{R}^{p+m}$ onto $E_m = \mathbb{R}^p \times \mathbb{T}^m$, and ν_t^m be a Gaussian measure on $\mathbb{R}^{p+m}$ such that

$$\widehat{\nu}_t^m(\theta) = \exp\{-t(A_m\theta \cdot \theta - ib^m\theta)\}, \quad \theta \in \mathbb{R}^{p+m}.$$

It is clear that the measure μ_t^m is the image of measure ν_t^m under the mapping γ_m. From the absolute continuity of μ_t^m it is easy to deduce the absolute continuity of ν_t^m. In particular, $\operatorname{supp}\nu_t^m = \mathbb{R}^{p+m}$. Therefore, in turn, it follows that $\operatorname{supp}\mu_t^m = E_m$. In view of the arbitrariness of m, we conclude that $\operatorname{supp}\mu_t = E$. Consequently $\widetilde{\mu}_t > 0$ (dx-almost everywhere). From the relation $\widetilde{\mu}_t = \widetilde{\mu}_{t/2} * \widetilde{\mu}_{t/2}$, now we deduce that $\widetilde{\mu}_t(x) > 0$, $\forall x \in E$. $\qquad\square$

Proposition 4.2.2. *For every $t > 0$ the following assertions are equivalent:*
1) *measure μ_t is absolutely continuous with respect to Haar measure and has continuous density $\mu_t(x)$;*
2) $\widehat{\mu}_t(\theta) \in L_1(\widehat{E}; d\theta)$, *$d\theta$ is the Haar measure on $\widehat{E}$.*

Proof. In correspondence with relation (4.2.2), it is sufficient to assume that μ_t is symmetric. If it has continuous density $\mu_t(x)$, then $\mu_t(x)$ is a continuous positive definite function. Therefore 2) follows from Bochner's theorem. On the other hand, 2) $\Rightarrow$1) follows from the inverse Fourier transform theorem. $\qquad\square$

Corollary 4.2.1. *For some sequence $\{a_k : a_k > 0\}$ let the following inequality be satisfied:*

$$\sum_{k=1}^{\infty} a_k\theta_k^2 \leq A\theta \cdot \theta, \quad \theta = (\theta_k)_1^{\infty} \in \widehat{E}.$$

If the series $\sum_{k=1}^{\infty} e^{-a_k t}$ converges, then the measure μ_t is absolutely continuous with respect to Haar measure and has continuous density.

Proof. According to (4.2.1), $|\widehat{\mu}_t(\theta)| = \exp(-tA\theta \cdot \theta)$, and therefore for certain numbers $\alpha, \beta, \gamma > 0$ we will have:

$$\int_{\widehat{E}} |\widehat{\mu}_t(\theta)|\, d\theta = \int_{\mathbb{R}^p \times \mathbb{Z}^{(\infty)}} \exp(-tA\theta \cdot \theta)\, d\theta \le$$

$$\le \int_{\mathbb{R}^p} \exp(-\alpha t |x|^2)\, dx \int_{\mathbb{Z}^{(\infty)}} \exp\left(-t \sum_{k=1}^{\infty} a_k \theta_k^2\right) d\theta \le$$

$$\le \beta \prod_{k \ge 1} \left(1 + 2 \sum_{\theta \ge 1} \exp(-ta_k \theta^2)\right) \le$$

$$\le \beta \prod_{k \ge 1} (1 + \gamma e^{-ta_k}) < \infty.$$

$\square$

4.2.5

In the following, it will be convenient to consider the process X as the limit for the simpler processes X_n defined on groups $E_n = \mathbb{R}^p \times \mathbb{T}^n$. This can be done in the following manner. The transition function $p(t; x, \Gamma)$ of process X is invariant with respect to homomorphism $\pi_{n\infty} : E \to E_n$. Therefore Theorem 3.3.3 on phase space transformation can be applied, and we can form the process $X_n = \pi_{n\infty}(X)$ on group E_n. We note in passing that $X_m = \pi_{mn}(X_n)$ for $m < n$. Thus the sequence of processes $\{X_n\}$ is projective with respect to systems of mappings $\{\pi_{mn}\}$, and process X is its projective limit (see Sec. 1.3.4).

4.2.6

We shall fix $n < \infty$ and discuss the properties of sheaf $\mathcal{S}_n$ of superharmonic functions of process X_n. We note first that process X_n is space-homogeneous and has almost surely continuous trajectories. Its corresponding Gaussian semigroup of measures (μ_t^n) is associated with a semigroup (μ_t) of process X by the relation

$$\mu_t^n(\Gamma) = \mu_t(\pi_{n\infty}^{-1}(\Gamma)), \quad \Gamma \in \mathcal{B}(E_n).$$

From this and from (4.2.1) we conclude that X_n has the corresponding diffusion matrix $A_n = (a_{ij})_1^{n+p}$ and drift vector $b^n = (b_i)_1^{n+p}$, where $A = (a_{ij})_1^{\infty}$ and $b = (b_i)_1^{\infty}$ are the diffusion matrix and drift vector of X.

Let γ be the natural homomorphism of $\mathbb{R}^{p+n}$ onto $E_n = \mathbb{R}^p \times \mathbb{T}^n$ and let X_n^* be the space-homogeneous process with almost surely continuous trajectories on $\mathbb{R}^{p+n}$ associated with matrix A_n and vector b^n. By Theorem 3.3.3 on the phase space transformation, the processes X_n and $\gamma(X_n^*)$ are equivalent. Therefore, in particular, it follows that on twice continuously differentiable functions, the $\mathbb{C}_0$-generator $\mathcal{L}_{0n}$

of process X_n coincides with the differential operator

$$\mathcal{L}_n = \sum_{i,j=1}^{n+p} a_{ij}\partial_i\partial_j + \sum_{i=1}^{n+p} b_i\partial_i. \tag{4.2.3}$$

Using the diffusion character of process $X_n = \gamma(X_n^*)$, it is possible, using the results of [54: Chapter 13, Sec. 3], to verify the following proposition.

Proposition 4.2.3. *The following assertions are equivalent:*
1) $\operatorname{rang} A_n = \dim E_n (= p + n)$;
2) $X_n \in \mathcal{B}$ *(see Sec. 3.2)*;
3) $(E_n, \mathcal{H}_{\mathcal{G}_n})$ *is a Brelot harmonic group.*

Under the fulfillment of equivalent conditions 1) – 3), *every function* $u \in \mathcal{H}_{\mathcal{G}_n}(V)$ *is infinitely differentiable and satisfies the equation* $\mathcal{L}_n u = 0$ *on* V.

In conclusion of this section, we note several useful relations. Let $\operatorname{rang} A_n = p + n$; then for all $t > 0$ the measure μ_t^n is absolutely continuous with respect to Haar measure dx on E_n and has continuous density $\mu_t^n(x)$. For $x \in \mathbb{R}^{p+n}$ we set: $|x|_{A_n} := A_n^{-1}x \cdot x$ and $|A_n| := \det A_n$, then

$$\mu_t^n(x) = (2\pi)^n (4\pi t)^{-(p+n)/2} |A_n|^{-1/2} \sum_{\substack{\theta \in \\ \{0\} \times \mathbb{Z}^n \subset \mathbb{Z}^{p+n}}} \exp\left\{ -\frac{|x - b^n \cdot t - 2\pi\theta|_{A_n}}{4t} \right\}$$

$$\tag{4.2.4}$$

In the next two inequalities, $(\mu_t^{s;n})$ and $(r_{s;n}^\lambda)$ are the Gaussian semigroup and resolvent of symmetric process $X_{s;n} = \pi_{n\infty}(X_s)$ (see Sec. 4.2.3), and $\mu_t^{s;n}(x)$ and $r_{s;n}^\lambda(x)$ are their corresponding densities with respect to Haar measure and $\Lambda_n := |b^n|_{A_n}$:

$$\frac{1}{2}\{\mu_t^n(x) + \mu_t^n(-x)\} \geq e^{-\Lambda_n t} \mu_t^{s;n}(x); \tag{4.2.5}$$

$$\frac{1}{2}\{r_n^\lambda(x) + r_n^\lambda(-x)\} \geq r_{s;n}^{\lambda+\Lambda_n}(x). \tag{4.2.6}$$

4.2.7

In this section we investigate the properties of sheaf $\mathcal{S}$ of superharmonic functions of process X, assuming that X is transient (see Sec. 4.2.2). For such an assumption the λ-potential kernel $R^\lambda(x, \Gamma)$ is defined also for $\lambda = 0$ and takes the form $R(x, \Gamma) = r(x - \Gamma)$.

Theorem 4.2.1. *The following assertions are equivalent:*
1) $(E, \mathcal{S})$ *is a* $\mathcal{P}$-*harmonic group;*
2) $(E, \mathcal{H}_{\mathcal{S}})$ *is a Bauer* $\mathcal{P}$-*harmonic group;*

3) *the measure $r = \int_0^\infty \mu_t\, dt$ is absolutely continuous with respect to Haar measure.*

If 1) – 3) are satisfied, then the following condition is satisfied:
4) $\forall n \geq 1$, *we have* $\mathrm{rang}\, A_n \geq \dim E_n - 1$.

Proof. (2) $\Rightarrow$1)): See [46: Corollary 3.1.1].

(1) $\Rightarrow$3)): For any function $\varphi \in \mathbb{B}_k^+$, the function $R\varphi = r * \varphi$ is excessive, and consequently, lower-semicontinuous; therefore 3) easily follows.

(3)$\Rightarrow$1) + 4): Let function $r(x)$ be the excessive density of measure r with respect to Haar measure. For any function $\varphi \in \mathbb{B}_k$, the function $R\varphi = r * \varphi$ is continuous. Now applying the theorem of Bliedtner and Hansen (Theorem 2.8.2), we conclude that $\mathcal{S}$ is a hyperharmonic sheaf and $(E, \mathcal{S})$ is a $\mathcal{P}$-harmonic group. We shall show that 4) is satisfied. Let r_n be the potential kernel associated with process $X_n = \pi_{n\infty}(X)$ (see Sec. 4.2.5) on group $E_n = \mathbb{R}^p \times \mathbb{T}^n$. The absolute continuity of r implies the absolute continuity of measure r_n. Let $\gamma : \mathbb{R}^{p+n} \to \mathbb{R}^p \times \mathbb{T}^n$ be the natural homomorphism, X_n^* be the space-homogeneous process on group $\mathbb{R}^{p+n}$ such that $X_n = \gamma(X_n^*)$ (see Sec. 4.2.6) and r_n^* be the potential kernel of process X_n^*. From the relation

$$r_n(\Gamma) = \sum_{\theta \in \{0\} \times \mathbb{Z}^n \subset \mathbb{Z}^{p+n}} r_n^*(\Gamma + 2\pi\theta)$$

it follows that the measure r_n^* is also absolutely continuous. Now let (ν_t^n) be the Gaussian semigroup associated with process X_n^*. We have

$$\widehat{\nu}_t^n(\theta) = \exp\{-t(A_n\theta \cdot \theta - ib^n\theta)\}, \ \theta \in \mathbb{R}^{p+n}.$$

If $\mathrm{rang}\, A_n < p + n - 1$, then for all $t > 0$ the measure ν_t^n is concentrated on a hypersurface $\prod$ such that $\dim \prod < p+n$. Consequently the measure $r_n^* = \int_0^\infty \nu_t^n\, dt$ is also concentrated on this hypersurface and thus is singular with respect to Haar measure ($=$ Lebesgue measure) - contradiction.

(3) $\Rightarrow$2)): Thus, it has been proved that $(E, \mathcal{S})$ is a $\mathcal{P}$-harmonic group. In order to prove 2), it remains to show that the family of regular sets forms the base of a topology [46: Corollary 3.1.2]. Let $\mathcal{S}_n$ be the sheaf of superharmonic functions of process X_n; then $(E_n, \mathcal{S}_n)$ is a $\mathcal{P}$-harmonic group and if the set $V \subset E_n$ is regular (with respect to $\mathcal{S}_n$), then the set $\pi_{n\infty}^{-1}(V) \subset E$ is regular (with respect to $\mathcal{S}$). Thus it is sufficient to show that for any $n \geq 1$, the family of regular sets in $(E_n, \mathcal{S}_n)$ forms the base of a topology. Finally, let $\mathcal{S}_n^*$ be the sheaf of superharmonic functions of process X_n^*; then $(\mathbb{R}^{p+n}, \mathcal{S}_n^*)$ is a $\mathcal{P}$-harmonic group and $\mathcal{S}_n \cdot \gamma \subset \mathcal{S}_n^*$. Since the mapping γ is locally an isomorphism, it is sufficient to show that the family of regular sets with respect to $\mathcal{S}_n^*$ forms the base of a topology. According to the previous section, there are two possibilities: a) $\mathrm{rang}\, A_n = n + p$, b) $\mathrm{rang}\, A_n = n + p - 1$. In case a) the family of regular sets forms the base of a topology by Proposition 4.2.3. In case b), performing a nondegenerate linear transformation, we arrive at a space-time Wiener process on group $\mathbb{R}^{p+n}$ and the sheaf

of solutions of the heat equation corresponding to this process. This sheaf, as it is well-known (see [53: Chapter XV]), has a base of regular sets, and consequently the original sheaf also has one.   □

We shall call a process X non-degenerate if the following condition is satisfied:[1]

$(\mathbb{R})$: $\forall n \geq 1$, rang $A_n = \dim E_n$.

Theorem 4.2.2. *The following assertions are equivalent:*
1) $(E, \mathcal{S})$ *is an elliptic $\mathcal{P}$-harmonic group;*
2) *the process X is non-degenerate and measure $r = \int_0^\infty \mu_t \, dt$ is absolutely continuous with respect to Haar measure.*

Proof. 1)$\Rightarrow$2): By Theorem 4.2.1 it is sufficient to show that X is non-degenerate. From the ellipticity of $(E, \mathcal{S})$ it follows that for every $n > 0$ the group $(E_n, \mathcal{S}_n)$ is elliptic. By Theorem 4.1.2, $(E_n, \mathcal{H}_{\mathcal{S}_n})$ is a Brelot harmonic group, and by Proposition 4.2.3, rang $A_n = n + p$.

2) $\Rightarrow$ 1): By Theorem 4.2.1, $(E, \mathcal{S})$ is a $\mathcal{P}$-harmonic group, and by Proposition 4.2.3, for every $n > 0$ the harmonic group $(E_n, \mathcal{S}_n)$ is elliptic. By Proposition 3.3.2, the sequence $\{E_n, \mathcal{S}_n; \pi_{mn}\}$ is projective and $(E, \mathcal{S})$ is its projective limit. Now the ellipticity of $(E, \mathcal{S})$ follows from the fact that every function $u \in \mathcal{S}^+(V)$, $V = \pi_{k\infty}^{-1}(V_k)$, is the limit of a monotonically increasing sequence of functions $u_n \cdot \pi_{n\infty}$ such that $u_n \in \mathcal{S}_n^+(V_n)$, $V_n = \pi_{kn}^{-1}(V_k)$ (see Sec. 3.4.12).   □

Theorem 4.2.3. *The following assertions are equivalent:*
1) $(E, \mathcal{H}_{\mathcal{S}})$ *is a Brelot $\mathcal{P}$-harmonic group;*
2) *the process X is non-degenerate and measure $r = \int_0^\infty \mu_t \, dt$ is absolutely continuous with respect to Haar measure, and its excessive density is continuous on the set $E \backslash \{0\}$.*

Proof. 1) $\Rightarrow$ 2): In correspondence with Theorem 4.2.2, it remains to prove that the function $r(x)$, an excessive density of measure r, is continuous on $E \backslash \{0\}$. We choose an open set $V \subset E$ such that $\overline{V} \cap \{0\} = \emptyset$ and we write the Hunt identity (see Sec. 2.10)

$$P_{E \backslash V}(x, r(\cdot, y)) = \widehat{P}_{E \backslash V}(y, r(x, \cdot)),$$

where $r(x, y) := r(x - y)$ is the Green's function of dual processes X and $\widehat{X}$. Setting $y = 0$, we obtain for any $x \in V$

$$r(x) = P_{E \backslash V}(x, r) = \sup_n P_{E \backslash V}(x, r \wedge n).$$

---

1   It is easy to see that this condition is equivalent to the fact that differential operator $\mathcal{L}$ associated with process X (see Sec. 4.2.1) is elliptic (see Sec. 4.1.3).

For every $n > 0$ the function $u_n = P_{E\setminus V}(\cdot, r \wedge n)$ is harmonic on V, and the sequence $\{u_n\}$ monotonically increases for $n \uparrow \infty$ and $\sup_n u_n = r < \infty$ (dx a.e. on V). By the Brelot convergence property, the function $r = \sup_n u_n$ is harmonic on V. In view of the arbitrariness of $V \subset E\setminus\{0\}$, we conclude, in particular, that the function $r(x)$ is continuous on $E\setminus\{0\}$.

2) $\Rightarrow$ 1): According to Theorem 4.2.2, $(E, \mathcal{H}_{\mathscr{S}})$ is an elliptic $\mathscr{P}$-harmonic Bauer group and thus it remains to show that the Brelot convergence property is satisfied. Let V be a connected open subset of E and $r_V(x, y)$ be the Green's function for the part of processes X and $\widehat{X}$ on V. This function is continuous on $V \times V\setminus\{(x, y): x = y\}$ and by the ellipticity property is strictly positive on $V \times V$. Arguing here as in Proposition 3.1.4[2], we arrive at the Harnack inequality, which implies the Brelot convergence property. $\qquad\square$

4.2.8

We set $\dot{E} = E \times \mathbb{R}^1$, and let $\dot{X}$ be the space-time process on group $\dot{E}$ associated with process X (see Sec. 3.2). We investigate the properties of sheaf $\dot{\mathscr{S}}$ of superharmonic functions of process $\dot{X}$. First we note that $\dot{X}$ is space-homogeneous and has almost surely continuous trajectories $t \rightarrow \dot{x}_t = (x_t, \tau - t)$ $(\dot{x}_0 = (x, \tau))$. Its corresponding semigroup and resolvent have the form

$$\dot{\mu}_t = \mu_t \otimes \varepsilon_t, \quad \dot{r}^\lambda = e^{-\lambda t} \mu_t 1_{\{t>0\}} dt.$$

From these relations it follows that on smooth cylindrical functions, the $\mathbb{C}_0$-generator $\dot{\mathscr{L}}_0$ of process $\dot{X}$ coincides with the infinite-dimensional parabolic differential operator

$$\dot{\mathscr{L}} = \mathscr{L} - \partial_t,$$

where $\mathscr{L}$ is the infinite-dimensional elliptic operator associated with process X (see Sec. 4.2.1).

The method of investigation of sheaf $\dot{\mathscr{S}}$ is similar to that which we applied for investigation of sheaf $\mathscr{S}$. As in Sec. 4.2.5, we have $\dot{X} = \lim \text{proj}\, \dot{X}_n$, where $\dot{X}_n$ is the space-time process on group $\dot{E}_n = E_n \times \mathbb{R}^1$ associated with process X_n. For every $n > 0$, $(\dot{E}_n, \mathcal{H}_{\dot{\mathscr{S}}_n})$ is a harmonic Bauer group with the Doob convergence axiom if and only if $\text{rang}\, \dot{A}_n = \dim \dot{E}_n - 1$. In this connection every function $u \in \mathcal{H}_{\dot{\mathscr{S}}_n}(V)$ is infinitely differentiable and satisfies the equation $\dot{\mathscr{L}}_n u = 0$ on V. Assuming that the condition

$$\forall n > 0 \quad \text{rang}\, \dot{A}_n = \dim \dot{E}_n - 1 \tag{$\dot{\mathbb{R}}$}$$

is satisfied, as in Sec. 4.2.7 we arrive at the "parabolic" variant of Theorems 4.2.2 and 4.2.3.

---

2   In the general case, $r(0) \neq +\infty$ and consequently $X \notin \mathscr{A}$ (see later Example 4.3.5).

Theorem 4.2.4. *The following assertions are equivalent:*
1) *$(\dot{E}, \mathcal{H}_{\dot{\varphi}})$ is a Bauer $\mathcal{P}$-harmonic group;*
2) *for every $t > 0$ the measure μ_t is absolutely continuous with respect to Haar measure.*

We note that the potential kernel $\dot{r}$ of process $\dot{X}$ has the form

$$\dot{r}(d\dot{x}) = \mu_t(dx)1_{\{t>0\}}dt, \quad \dot{x} = (x, t).$$

Therefore, if the equivalent assertions of Theorem 4.2.4 hold, then the excessive density $\dot{r}(\dot{x})$ of measure $\dot{r}$ takes the form

$$\dot{r}(\dot{x}) = \begin{cases} \mu_t(x), & t > 0 \\ 0, & t \le 0. \end{cases}$$

Theorem 4.2.5. *The following assertions are equivalent:*
1) *$(\dot{E}, \mathcal{H}_{\dot{\varphi}})$ is a Bauer $\mathcal{P}$-harmonic group with the Doob convergence axiom;*
2) *the function $\dot{r}(\dot{x})$ is continuous on the set $\dot{E}\backslash\{\dot{0}\}$.*

Remark 4.2.1. It is easy to show that the potential kernels r and $\dot{r}$ of processes X and $\dot{X}$, as generalized functions, satisfy the equations

$$\mathcal{L}u = -\varepsilon_0, \quad \dot{\mathcal{L}}u = -\varepsilon_{\dot{0}}$$

Thus, measures r and $\dot{r}$ are "fundamental solutions" for the elliptic and parabolic infinite-dimensional differential operators $\mathcal{L}$ and $\dot{\mathcal{L}}$ respectively. The results of Secs. 4.2.7 and 4.2.8 now mean that the properties of weak solutions of homogeneous equations are equivalent to the corresponding properties of fundamental solutions. In the finite-dimensional case the fundamental solutions r and $\dot{r}$ simultaneously have the required properties. We give the corresponding infinite-dimensional result.

Theorem 4.2.6. *Let condition $(\mathbb{C})$ be satisfied (see Sec. 4.2.4); then the following assertions are equivalent:*
1) *$(E, \mathcal{H}_{\varphi})$ is an elliptic Bauer $\mathcal{P}$-harmonic group;*
2) *$(\dot{E}, \mathcal{H}_{\dot{\varphi}})$ is a Bauer $\mathcal{P}$-harmonic group.*

Proof. According to Theorems 4.2.2 and 4.2.4, assertion 1) is equivalent to absolute continuity of the resolvent (r^λ), and assertion 2) is equivalent to absolute continuity of semigroup (μ_t) with respect to Haar measure. Now it remains to use Proposition 4.2.1. $\qquad\qquad\square$

Remark 4.2.2. It is not difficult to verify that the results of this section carry over word for word to the case of a process X^λ which is the λ-subprocess of process X.

4.3 Space homogeneous processes and harmonic functions: quasidiagonal case

In this section we continue the discussion of properties of sheaves $\mathcal{S}$ and $\dot{\mathcal{S}}$ of superharmonic functions of processes X and $\dot{X}$. First, we give a theorem similar to Theorem 4.2.6, but concerning more rigid axiomatic properties. And second, we state these properties not in terms of fundamental solutions r and $\dot{r}$ of differential operators $\mathcal{L}$ and $\dot{\mathcal{L}}$, but directly in terms of the coefficients $A = (a_{ij})_1^\infty$ and $b = (b_i)_1^\infty$ of these operators.

4.3.1

For the entire section it is assumed that along with condition ($\mathbb{R}$) (see Sec. 4.2.7), the following conditions are satisfied:

(Λ) $\Lambda := \sup_n A_n^{-1} b^n \cdot b^n < \infty$;

(D) $A = \mathrm{diag}(\mathcal{A}_k)_1^\infty$, $\dim \mathcal{A}_k < \infty$ $(k = 1, 2, ...)$;

we shall call a process X with such a diffusion matrix *quasidiagonal*.

   We shall make several remarks.

1. It is not difficult to prove that (Λ) $\Rightarrow$ ($\mathbb{C}$) (see Sec. 4.2.4). The fact that ($\mathbb{C}$) $\nRightarrow$ (Λ) is shown by the example: $A = \mathrm{diag}(a_k)_1^\infty$, $b = (b_k)_1^\infty$: $a_k = k^2$, $b_k = k$ $(k = 1, 2, ...)$.
2. From condition (D) it follows that for every $t > 0$ the Gaussian measure μ_t associated with matrix tA and vector tb has the form

$$\mu_t = \otimes_{k=1}^\infty \mu_t^k, \tag{4.3.1}$$

where μ_t^k is the Gaussian measure on finite-dimensional group $\mathcal{E}_k$ associated with matrix $t\mathcal{A}_k$ and vector tb^k such that $E = \prod_{k=1}^\infty \mathcal{E}_k$, $A = \mathrm{diag}(\mathcal{A}_k)_1^\infty$, and $b = (b^k)_1^\infty$.

   Of the three given conditions, ($\mathbb{R}$), (Λ), and (D), the most restrictive is condition (D). In the following section we shall try to eliminate it.

4.3.2

Let X_s be the symmetric process associated with process X (see Sec. 4.2.3) and $\mathcal{S}_s$ be the sheaf of superharmonic functions of process X_s.

Theorem 4.3.1. *The following assertions are equivalent:*
1) *$(E, \mathcal{H}_{\mathcal{S}})$ is a Brelot harmonic group;*
2) *$(\dot{E}, \mathcal{H}_{\dot{\mathcal{S}}})$ is a Bauer $\mathcal{P}$-harmonic group with the Doob convergence property;*
3) *$(E, \mathcal{H}_{\mathcal{S}_s})$ is a Brelot harmonic group;*
4) *$(\dot{E}, \mathcal{H}_{\dot{\mathcal{S}}_s})$ is a Bauer $\mathcal{P}$-harmonic group with the Doob convergence property.*

We shall first prove two lemmas.

Lemma 4.3.1. *Let X be a symmetric process. We assume that the equivalent asser-tions of Proposition 4.2.1 hold. If for some $\lambda = \lambda_0$ the function $r^{\lambda_0}(x)$ is continuous on $E\backslash\{0\}$, then:*
1) *the function $\mu_t(x)$ is continuous on E for all $t > 0$;*
2) *the function $r^\lambda(x)$ is continuous on $E\backslash\{0\}$ for all $\lambda > 0$ (in the transient case, $\lambda \geq 0$).*

Proof. For $x = (x_i)_1^\infty \in E = \prod_{i=1}^\infty \mathscr{E}_i$ such that $x_1 \neq 0$ and $x_i = 0$ for $i > 1$, we shall have

$$\int_0^\infty e^{-\lambda_0 t} \mu_t^1(x_1) \prod_{i \neq 1} \mu_t^i(0)\, dt = \int_0^\infty e^{-\lambda_0 t} \mu_t(x)\, dt = r^{\lambda_0}(x) < \infty,$$

which implies that $\prod_{i \neq 1} \mu_t^i(0) < \infty$, and also $\mu_t(0) < \infty$ (a.e. dt). The function $t \to \mu_t(0)$ monotonically decreases, therefore $\mu_t(0) < \infty$ for all $t > 0$. Further, $\mu_t(x) \leq \mu_t(0)$; from the relation $\mu_t = \mu_\Delta * \mu_{t-\Delta}$ we now get the continuity of function $\mu_t(x)$.

The continuity of $r^\lambda(x)$ for $\lambda \geq \lambda_0$ follows from the obvious inequality $r^\lambda \leq r^{\lambda_0}$ and the inclusion $r^\lambda \in \mathscr{H}_{\mathscr{G}\lambda}(E\backslash\{0\})$. For $\lambda : 0 < \lambda < \lambda_0$ we have

$$r^\lambda = \int_0^\infty e^{-\lambda t} \mu_t\, dt = \int_0^1 + \int_1^\infty \leq e^{(\lambda_0 - \lambda)} r^{\lambda_0} + \mu_1(0)/\lambda.$$

If the process X is transient, then $E = \mathbb{R}^p \times \mathbb{T}^\infty$ $(p \geq 3)$ (see Sec. 4.2.2). Without loss of generality, we may assume that (μ_t^1) is a Gaussian semigroup on $\mathbb{R}^p$. We have

$$r \leq e^{\lambda_0} r^{\lambda_0} + \int_1^\infty \mu_t(0)\, dt \leq e^{\lambda_0} r^{\lambda_0} + \frac{\mu_1(0)}{\mu_1^1(0)} \cdot \int_1^\infty \mu_t^1(0)\, dt.$$

Thus in all cases we get that r^λ is locally bounded on $E\backslash\{0\}$. The inclusion $r^\lambda \in \mathscr{H}_{\mathscr{G}\lambda}(E\backslash\{0\})$ now gives its continuity. $\qquad\square$

Lemma 4.3.2. *For the same assumptions as in Lemma 4.3.1, the following con-ditions are equivalent:*
1) *the function $r^\lambda(x)$ is continuous on $E\backslash\{0\}$ for all $\lambda > 0$ (in the transient case, $\lambda \geq 0$);*
2) *the function $\dot{r}(\dot{x}) = \begin{cases} \mu_t(x), & t > 0 \\ 0, & t \leq 0 \end{cases}$ is continuous on $\dot{E}\backslash\{\dot{0}\}$;*
3) *for all $c > 0$, we have $\lim_{t \to 0} e^{-c/t} \mu_t(0) = 0$;*
4) *the function $q_t(x)$ (see Sec. 4.2.4) is continuous on E for all $t > 0$;*
5) *$\widehat{q}_t(\theta) = \exp(-t\sqrt{A\theta \cdot \theta}) \in L_1(\widehat{E}; d\theta)$ for all $t > 0$.*

Proof. 1) $\Rightarrow$ 3): We choose $x_1 \in \mathscr{E}_1$ such that $\frac{1}{4}\mathscr{A}_1^{-1}x_1 \cdot x_1 = c$; then for $x = (x_1, 0, ...)$ and $\lambda > 0$ we will have:

$$\int_0^1 e^{-c/t}\mu_t(0)\,dt = \int_0^1 e^{-c/t}\mu_t^1(0)\prod_{i\neq 1}\mu_t^i(0)\,dt \le$$

$$\le \int_0^1 \mu_t^1(x_1)\prod_{i\neq 1}\mu_t^i(0)\,dt \le e^\lambda \int_0^1 e^{-\lambda t}\mu_t(x)\,dt \le e^\lambda r^\lambda(x) < \infty.$$

Using the inverse Fourier transform, we write the equality

$$\mu_t(0) = \int_{\widehat{E}}\exp(-tA\theta \cdot \theta)\,d\theta.$$

Differentiating this equality with respect to t and considering that $ae^{-a/2} < 1$, $(a > 0)$ we get $|\partial_t\mu_t(0)| \le \frac{1}{t}\mu_{t/2}(0)$. With the help of this inequality we get that, as above,

$$\int_0^1 e^{-c/t}|\partial_t\mu_t(0)|\,dt < \infty.$$

Now integrating by parts, we obtain 3).

3) $\Rightarrow$ 2): The function $\dot{r}(\dot{x})$ is continuous on each of the semispaces $\dot{E}^+ = E \times \{t > 0\}$ and $\dot{E}^- = E \times \{t < 0\}$, therefore it remains to show its continuity at points of the set $E \times \{t = 0\}\setminus\{\dot{0}\}$. Let $\dot{x} = (x, t) \in \dot{E}^+$ and $\dot{x} \to \dot{y} = (y, 0) \neq \dot{0}$; we shall show that $\dot{r}(\dot{x}) \to 0 = \dot{r}(\dot{y})$. We shall assume that $y_1 \neq 0$ $(y = (y_i)_1^\infty)$. Then, since $x_1 \to y_1$, we can assume that for some $\nu > 0$ we have $\frac{1}{4}\mathscr{A}_1^{-1}x_1 \cdot x_1 \ge \nu$. We write the inequality

$$\dot{r}(\dot{x}) = \mu_t(x) = \mu_t^1(x_1)\prod_{i\neq 1}\mu_t^i(x_i) \le$$

$$\le \mu_t^1(x_1)\prod_{i\neq 1}\mu_t^i(0) = \mu_t^1(x_1)/\mu_t^1(0) \cdot \mu_t(0).$$

From (4.2.4) the following estimate may be easily obtained: there exist numbers $\alpha, \beta \ge 1$ such that $\forall t \in \,]0, 1]$, $\forall x_1 \in \mathscr{E}_1$ we have

$$\mu_t^1(x_1) \le \alpha\mu_t^1(0)\exp\{-\mathscr{A}_1^{-1}x_1 \cdot x_1/4\beta t\}.$$

Applying this estimate and setting $c = \nu/\beta$ we get

$$\dot{r}(\dot{x}) \le \alpha e^{-c/t}\mu_t(0), \ \ 0 < t \le 1;$$

from which, according to 3), it follows that $\dot{r}(\dot{x}) \to 0$ for $\dot{x} \to \dot{y} = (y, 0)$.

2) $\Rightarrow$ 1): According to Remark 4.2.2 and Theorem 4.2.5, $(\dot{E}, \mathscr{H}_{\dot{\varphi}\lambda})$ is a Bauer harmonic group with the Doob convergence axiom. The function $u(\dot{x}) := r^\lambda(x)$ is superharmonic, and in addition the Hunt identity implies that $\forall \dot{V} \subset \dot{E} : \bar{V} \cap (\{0\} \times \mathbb{R}^1) = \emptyset$

$$\dot{P}^{\lambda}_{\dot{E}\backslash\dot{V}}(\dot{x}, u) = u(\dot{x}), \quad \dot{x} \in \dot{V}.$$

We set $u_n := \dot{P}^{\lambda}_{\dot{E}\backslash\dot{V}}(\cdot, u \wedge n)$; then for all $n > 0$ we have $u_n \in \mathcal{H}_{\mathcal{G}\lambda}(\dot{V})$, $u_n \uparrow u < \infty$ (a.e. $d\dot{x}$ on $\dot{V}$). By the Doob convergence axiom we have $u \in \mathcal{H}_{\mathcal{G}\lambda}(\dot{V})$, and in particular we get that $u \in \mathbb{C}(\dot{E}\backslash(\{0\} \times \mathbb{R}^1))$, which implies $r^{\lambda} \in \mathbb{C}(E\backslash\{0\})$.

4) $\Leftrightarrow$ 5): The proof is carried out in the same way as in Proposition 4.2.2.

3) $\Rightarrow$ 4): For fixed $t > 0$, the function $\mu_s(x)h_t(s)$ (see Sec. 4.2.4) is bounded by $s \in \,]0, 1]$ and does not exceed $\text{const} \cdot s^{-3/2}$ for $s \in [1, \infty[$ uniformly for $x \in E$. Therefore the continuity of function $q_t(x) = \int_0^{\infty} \mu_s(x)h_t(s)\,ds$ follows from the continuity of $\mu_s(x)$ and the Lebesgue theorem.

4) $\Rightarrow$ 3): For any $c > 0$ and for $t^2 < 4c$ we shall have

$$\int_0^1 e^{-c/s}\mu_s(0)\,ds \leq \text{const} \int_0^1 \mu_s(0)h_t(s)\,ds < \text{const} \cdot q_t(0) < \infty.$$

Now the argument concludes as in part 1) $\Rightarrow$ 3). $\qquad\qquad\square$

Proof of Theorem 4.3.1. The implications 2) $\Rightarrow$ 1) and 4) $\Rightarrow$ 3) are true by Corollary 3.2.1. According to Theorems 4.2.3 and 4.2.5, each of the assertions of the theorem is equivalent to the corresponding property of the semigroup or the resolvent. Therefore the implication 2) $\Leftrightarrow$ 4) is true by relation (4.2.2); 3) $\Rightarrow$ 4) by Lemma 4.3.2, and finally, the truth of 1) $\Rightarrow$ 3) follows from inequality (4.2.6), condition (Λ), and Lemma 4.3.1. $\qquad\qquad\square$

Remark 4.3.1. Process X, for which the equivalent assertions of Theorem 4.3.1 hold, is in class $\mathcal{B}$ (see Sec. 3.2) (in the transient case $X \in \mathcal{A}$ (see Sec. 3.1)). This property is deduced from inequality (4.2.5) (respectively (4.2.6)) and condition (Λ). Conditions ($\mathbb{C}$) and (Λ) involved in Theorems 4.2.6 and 4.3.1 essentially enrich the structure of harmonic space $(E, \mathcal{H}_{\mathcal{G}})$. For example, from condition ($\mathbb{C}$) it follows that in $(E, \mathcal{H}_{\mathcal{G}})$ the domination principle, which is important in potential theory, is satisfied (see [46: Chapter 9]). From the more rigid condition (Λ) it is possible to deduce the strengthening of this principle, approximating it to the classical principle (see [45: Chapter 6, Sec. 4], [74: Chapter 1, § 3]).

4.3.3

Theorem 4.3.1 shows that the investigation of sheaf $\mathcal{G}$ reduces to the investigation of a simpler object - the sheaf $\mathcal{G}_s$ of superharmonic functions of symmetric process X_s. The next proposition (4.3.1) can prove to be useful in this sense.

Let $(a_k)_1^{\infty}$ be a successively indexed sequence of eigenvalues of matrices $\mathcal{A}_k$ $(k = 1, 2, \ldots)$ and X_d a symmetric process with diagonal diffusion matrix $A_d = \text{diag}(a_k)_1^{\infty}$. Let n_k be the dimension of matrix $\mathcal{A}_k$ and $||\mathcal{A}_k|| = \sup_{|\theta|=1} \mathcal{A}_k\theta \cdot \theta$.

Proposition 4.3.1. *If one of the following conditions is satisfied:*
1) $n := \sup_k n_k < \infty$;
2) $m := \sup_k \|\mathscr{A}_k\| \, \|\mathscr{A}_k^{-1}\| < \infty$,

then the implication $X_d \in \mathscr{B} \Rightarrow X \in \mathscr{B}$ is true (see Sec. 3.2).

Proof. Obviously it is sufficient to consider the case $E = \mathbb{T}^\infty$ and X a symmetric process on E. Let condition 1) be satisfied, ν_k the largest eigenvalue of matrix $\mathscr{A}_k$, and $\mathscr{A}_k^d$ the diagonal matrix consisting of eigenvalues of matrix $\mathscr{A}_k$. We have

$$
\mu_t(0) = \int_{\mathbb{Z}^{(\infty)}} e^{-tA\theta\cdot\theta}\,d\theta = \prod_{k\geq 1}\int_{\mathbb{Z}^{n_k}} e^{-t\mathscr{A}_k\theta\cdot\theta}\,d\theta \leq
$$

$$
\leq \prod_{k\geq 1}\int_{\mathbb{Z}^{n_k}} e^{-t\nu_k|\theta|^2}\,d\theta = \prod_{k\geq 1}\left(\int_{\mathbb{Z}^1} e^{-t\nu_k\theta^2}\,d\theta\right)^{n_k} \leq
$$

$$
\leq \prod_{k\geq 1}\left(\int_{\mathbb{Z}^1} e^{-t\nu_k\theta^2}\,d\theta\right)^n \leq \prod_{k\geq 1}\left(\int_{\mathbb{Z}^{n_k}} e^{-t\mathscr{A}_k^d\theta\cdot\theta}\,d\theta\right)^n =
$$

$$
= \left(\int_{\mathbb{Z}^{(\infty)}} e^{-tA_d\theta\cdot\theta}\,d\theta\right)^n = (\mu_t^d(0))^n,
$$

where (μ_t^d) is the Gaussian semigroup associated with process X_d. The obtained inequality together with Lemma 4.3.2. 3) shows that the mentioned implication is true.

If condition 2) is satisfied, then arguing as above, we arrive at the inequality $\mu_t(0) \leq \mu_{t/m}^d(0)$, and then we use Lemma 4.3.2 3).   □

4.3.4

In this section we consider the symmetric process X with diagonal diffusion matrix $A = \mathrm{diag}(a_k)_1^\infty$. In addition, we assume that $E = \mathbb{T}^\infty$. According to Sec. 4.2.2, this assumption does not limit generality. In correspondence with (4.3.1) we have

$$
\mu_t = \otimes_{k=1}^\infty \mu_t^k, \tag{4.3.2}
$$

where μ_t^k is the Gaussian measure on $\mathscr{C}_k = \mathbb{T}^1$ with density $\mu_t^k(x_k)$ with respect to Haar measure dx_k on $\mathscr{C}_k$:

$$
\mu_t^k(x_k) = 2\pi(4\pi a_k t)^{-1/2} \sum_{n=-\infty}^{\infty} \exp\left\{ -\frac{(x_k - 2\pi n)^2}{4a_k t} \right\}. \tag{4.3.3}
$$

Denoting the Gaussian measure and density with $a. = 1$ by n_t and $n_t(x)$, we get $\mu_t^k = n_{a_k t}$, $\mu_t^k(x_k) = n_{a_k t}(x_k)$ and $\mu_t = \otimes_{k=1}^\infty n_{a_k t}$.

For $t > 0$ we set $\varphi(t) = \sum_{k=1}^\infty \exp(-tk^2)$. If the measure μ_t is absolutely continuous with respect to Haar measure and has continuous density $\mu_t(x)$, then applying the inverse Fourier transform, we get

$$\mu_t(0) = \prod_{k=1}^{\infty}(1 + 2\varphi(a_k t)). \tag{4.3.4}$$

In the following we shall need some properties of function $\varphi(t)$ (see [30: 2.6]):

$$\varphi(t) \leq \frac{1}{2}\sqrt{\frac{\pi}{t}}; \tag{4.3.5}$$

$$\varphi(t) \sim \frac{1}{2}\sqrt{\frac{\pi}{t}}, \quad t \downarrow 0; \tag{4.3.6}$$

$$\varphi(t) \sim e^{-t}, \quad t \uparrow +\infty. \tag{4.3.7}$$

Theorem 4.3.2. *For every $t > 0$ the following assertions are true:*
1) *measure μ_t is absolutely continuous with respect to Haar measure if and only if the series $\sum_{k=1}^{\infty} e^{-2a_k t}$ converges;*
2) *the density $\mu_t(x)$ of measure μ_t is continuous if and only if the series $\sum_{k=1}^{\infty} e^{-a_k t}$ converges.*

We outline the proof of this theorem (see [30: Theorems 4.3 and 4.6]). We set $\rho(t) = \int_{T^1} \sqrt{n_t(x)}\, dx$. By Kakutani's theorem [96: Chapter 7, 6, Theorem 3] the absolute continuity of measure μ_t is equivalent to convergence of the infinite product $\prod_{k=1}^{\infty} \rho(a_k t)$. The investigation of this infinite product is based on the following properties of the function $\rho(t)$ [30: Proposition 2.8]:
1) $0 < \rho(t) < 1$,
2) $\lim_{t \to 0} \rho(t) = 0$,
3) $1 - \rho(t) \sim \frac{1}{4} e^{-2t}$, $t \uparrow \infty$.

The truth of assertion 2) follows from (4.3.4) and (4.3.7) (see [7: Theorem 4.1].$\square$

According to Remark 4.3.1, each assertion of Theorem 4.3.1 is equivalent to process X belonging to class $\mathscr{B}$ (see Sec. 3.2). In correspondence with Lemma 4.3.2 2) $\Rightarrow$ 3) the inclusion $X \in \mathscr{B}$ is expressed in the form of the following properties of Gaussian semigroup (μ_t) corresponding to process X:

$$\forall t > 0 \quad \mu_t \sim dx, \tag{4.3.8}$$

$$\forall \varepsilon > 0 \quad \lim_{t \downarrow 0} e^{-\varepsilon t} \mu_t(0) = 0. \tag{4.3.9}$$

The proof of the following theorem, which is the principal theorem in this section, in essence reduces to a proof of the equivalence of these properties to some asymptotic properties of sequence (a_k) corresponding to (μ_t).

For $\lambda > 0$ we define a function

$$N(\lambda) := \#\{k \geq 1 : a_k \leq \lambda\}.$$

Theorem 4.3.3. *The following assertions are equivalent:*
1) *$(E, \mathcal{H}_{\mathscr{G}})$ is a Brelot harmonic group,*
2) *$(\dot{E}, \mathcal{H}_{\dot{\mathscr{G}}})$ is a Bauer $\mathcal{P}$-harmonic group with the Doob convergence property,*
3) *$N(\lambda) = o(\lambda)$, $\lambda \uparrow \infty$.*

Proof. As noted above, the proof of the theorem reduces to a proof of the equivalence of properties (4.3.8)–(4.3.9) to property 3). According to Theorem 4.3.2, property (4.3.8) is equivalent to the convergence of series $\sum_1^\infty e^{-\varepsilon a_k}$ for any $\varepsilon > 0$. We write the identity

$$\sum_1^\infty e^{-\varepsilon a_k} = \int_0^\infty e^{-\varepsilon x} dN(x),$$

from which we easily deduce that the convergence of series $\sum_1^\infty e^{-\varepsilon a_k}$ for all $\varepsilon > 0$ is equivalent to the following property of function $N(\lambda)$:

$$\log N(\lambda) = o(\lambda), \ \lambda \uparrow \infty. \tag{4.3.10}$$

Now we discuss property (4.3.9). The following chain of equalities holds:

$$\begin{aligned}
\log \mu_t(0) &= \sum_1^\infty \log(1 + 2\varphi(a_k t)) = \\
&= \int_0^\infty \log(1 + 2\varphi(xt)) \, dN(x) = \\
&= \int_0^\infty N\left(\frac{x}{t}\right) \frac{(-2\varphi'(x))}{1 + 2\varphi(x)} \, dx.
\end{aligned} \tag{4.3.11}$$

Now let property (4.3.9) be satisfied. Then for the given $\varepsilon > 0$ we choose $t_\varepsilon > 0$ such that for $0 < t < t_\varepsilon$ the inequality

$$\frac{\varepsilon}{t} \geq \log \mu_t(0) = \int_0^\infty N\left(\frac{x}{t}\right) \frac{(-2\varphi'(x))}{1 + 2\varphi(x)} \, dx$$

is satisfied. Now we note that $-\varphi'(x) = \sum_1^\infty k^2 e^{-k^2 x} \geq \varphi(x)$, and therefore

$$\frac{\varepsilon}{t} \geq \int_0^\infty N\left(\frac{x}{t}\right) \frac{2\varphi(x)}{1 + 2\varphi(x)} \, dx. \tag{4.3.12}$$

Considering (4.3.5)–(4.3.7), we estimate the integrals

$$\int_0^1 N\left(\frac{x}{t}\right) \frac{2\varphi(x)}{1 + 2\varphi(x)} \, dx \geq c_1 \int_0^1 e^{-x} N\left(\frac{x}{t}\right) \, dx,$$

$$\int_1^\infty N\left(\frac{x}{t}\right) \frac{2\varphi(x)}{1 + 2\varphi(x)} \, dx \geq c_2 \int_1^\infty e^{-x} N\left(\frac{x}{t}\right) \, dx.$$

Now we continue the inequality (4.3.12):

$$\frac{\varepsilon}{t} \geq c_3 \int_0^\infty N\left(\frac{x}{t}\right) e^{-x} \, dx \geq c_3 \int_1^\infty N\left(\frac{x}{t}\right) e^{-x} \, dx \geq c_4 N\left(\frac{1}{t}\right).$$

Setting $\frac{1}{t} = \lambda$ in this inequality, we get $N(\lambda) \le \varepsilon c_5 \lambda$, $\lambda > \lambda_\varepsilon$, where c_5 is an absolute constant, i.e. property 3) is satisfied.

Conversely, let property 3) be satisfied. For the given $\varepsilon > 0$, we choose $\lambda = \lambda_\varepsilon$ such that $N(\lambda) < \varepsilon \lambda$ for $\lambda > \lambda_\varepsilon$. With consideration of equality (4.3.11), we write $\log \mu_t(0)$ in the form of the sum of two integrals over intervals $]0, 1]$ and $[1, \infty[$ respectively. Let $t < 1/\lambda_\varepsilon$; then $\frac{x}{t} > \lambda_\varepsilon$ for $x \ge 1$ and consequently

$$\int_1^\infty N\left(\frac{x}{t}\right) \frac{(-2\varphi'(x))}{1 + 2\varphi(x)}\, dx \le \frac{\varepsilon}{t} \int_1^\infty \frac{-2x\varphi'(x)}{1 + 2\varphi(x)}\, dx = \varepsilon b_1 \cdot \frac{1}{t}.$$

For an estimate of the integral for interval $]0, 1]$, we note that by properties (4.3.5) and (4.3.6) of function $\varphi(x)$ and inequality $-\varphi'(x) \le \frac{2e^{-1}}{x}\varphi(\frac{x}{2})$, the function $\frac{-2\varphi'(x)}{1 + 2\varphi(x)}$ is dominated on this interval by the function b_2/x, where b_2 is a constant not dependent on ε. Thus

$$\int_0^1 N\left(\frac{x}{t}\right) \frac{(-2\varphi'(x))}{1 + 2\varphi(x)}\, dx \le b_2 \int_0^1 N\left(\frac{x}{t}\right) \frac{dx}{x} = b_2 \int_0^{1/t} \frac{N(y)}{y}\, dy.$$

Further, we have

$$\int_0^{1/t} \frac{N(y)}{y}\, dy = \int_0^{\lambda_\varepsilon} \frac{N(y)}{y}\, dy + \int_{\lambda_\varepsilon}^{1/t} \frac{N(y)}{y}\, dy \le$$

$$\le A_\varepsilon + \varepsilon \left(\frac{1}{t} - \lambda_\varepsilon\right) = \frac{1}{t}\left[tA_\varepsilon + \varepsilon(1 - t\lambda_\varepsilon)\right].$$

Thus finally we get the inequality

$$t \log \mu_t(0) \le \varepsilon b_1 + t b_2 A_\varepsilon + \varepsilon b_2(1 - t\lambda_\varepsilon)$$

which is satisfied for $t < 1/\lambda_\varepsilon$. From the obtained inequality we conclude that

$$\limsup_{t \downarrow 0} t \log \mu_t(0) \le \varepsilon(b_1 + b_2).$$

In view of the arbitrariness of $\varepsilon > 0$, we conclude that property (4.3.9) is satisfied. $\qquad\square$

Corollary 4.3.1. *The following two conditions are sufficient for the truth of assertions 1–3 of Theorem* 4.3.3:
(N_1) $\lim_{n\to\infty} a_n/n = \infty$,

(N_1^*) $\sum_1^\infty \frac{1}{a_n} < \infty$.

For completeness we give the following result.

Theorem 4.3.4. *The following assertions are equivalent:*
1) *$(E, \mathcal{H}_\varphi)$ is a Bauer elliptic harmonic group;*
2) *$(\dot{E}, \mathcal{H}_\varphi)$ is a Bauer $\mathcal{P}$-harmonic group,*
3) *$\log N(\lambda) = o(\lambda)$, $\lambda \uparrow \infty$.*

Proof. According to Theorems 4.2.5 and 4.2.6, assertions 1) and 2) are equivalent to the absolute continuity of measure μ_t with respect to Haar measure for any $t > 0$. The last property, as was shown in the first part of the proof of Theorem 4.3.3, is equivalent to property 3). $\square$

4.3.5

Example. Let $E = \mathbb{R}^3 \times \mathbb{T}^\infty$ and let X be the space-homogeneous process on E defined by diffusion matrix $A = \mathrm{diag}(a_k)_1^\infty$, with $a_k = k^{2+\delta}$, $\delta > 0$ and drift vector $b = (b_k)_1^\infty$ such that $b_k = k^{2+3\delta}$. We have:

$$\Lambda = \sum_{k=1}^\infty b_k^2/a_k = \infty,$$

and thus condition (Λ) (see Sec. 4.3.1) is not satisfied in this example. In addition, according to Theorem 4.3.3, semigroup (μ_t^s) and resolvent (r_s^λ) (see Sec. 4.2.3) satisfy the conditions of Theorems 4.2.3 and 4.2.5, and semigroup (μ_t) (see (4.2.2)) and resolvent (r^λ) also satisfy these conditions. Thus, all assertions of Theorem 4.3.1 are satisfied for process X. We shall show that:
1) $r(0) < \infty$,
2) $\lim_{t \to 0} \mu_t(0) = 0$.

Properties 1) and 2) mean that process X is not in class $\mathcal{A}$ or in class $\mathcal{B}$ (cf. Remark 4.3.1). In addition, each of the prelimit processes X_n (see Sec. 4.2.5) is in both class $\mathcal{A}$ and class $\mathcal{B}$ (see Proposition 4.2.3). This example is interesting also in other respects. For example, in Brelot space $(E, \mathcal{H}_{\mathcal{G}})$ every one-point set is polar, and the Green's function $r(x, y) = r(x - y)$ is finite on the diagonal (see [46: Remark 2, p. 282]). Property 1) also shows that the classical domination principle (see [74, Chapter 1, § 3]), [45: Chapter 6, Sec. 4]) does not hold in $(E, \mathcal{H}_{\mathcal{G}})$ (see also [46: Theorem 9.2.11 and Exercise 9.2.2] and [40: Theorem 4.12, Chapter 6 and remark, p. 292]).

It is obvious that 2) $\Rightarrow$ 1), therefore we shall prove 2). We have

$$\mu_t(0) = \mu_t^s(bt) = \prod_{k=1}^3 n_{a_k t}^*(b_k t) \prod_{k=4}^\infty n_{a_k t}(b_k t),$$

where $n_t^*(x)$ is the density of the normal distribution on $\mathbb{R}^1$ with parameters $(0, 2t)$, and $n_t(x)$ is the density of the normal distribution on $\mathbb{T}^1$ with the same parameters. For $x \in \mathbb{R}^1$ we denote by $\{x\}$ the number in interval $[-1, 1]$ such that $x - \{x\} = 2m$ ($m \in \mathbb{Z}$). Then

$$n_{a_k \pi t}(b_k \pi t) = n_{a_k \pi t}(\pi\{b_k t\}), \quad k \geq 4.$$

It is easy to verify the inequality

$$n_t(x) \le \left(1 + 2\sqrt{\frac{\pi}{t}}\right) \exp\left(-\frac{x^2}{4t}\right), \quad |x| \le \pi,$$

from which we get, for some number $\gamma > 0$ and any $0 < t < 1$,

$$\mu_{\pi t}^s(b\pi t) \le \exp\left\{-\frac{\pi}{4\sqrt{t}}\left(\frac{1}{\sqrt{t}}\sum_{k=1}^{\infty}\frac{\{b_k t\}^2}{a_k} - \gamma\right)\right\}.$$

Considering that $b_k^2/a_k = k^{2+5\delta}$, we get for $t \downarrow 0$

$$\sum_{k=1}^{\infty}\frac{\{b_k t\}^2}{a_k} \ge t^2 \sum_{k:b_k t \le 1}\frac{b_k^2}{a_k} \sim t^{1/2-\varepsilon} \quad \left(0 < \varepsilon < \frac{1}{6}\right).$$

The last two inequalities imply the truth of property 2). $\square$

Remark 4.3.2. The results of this section carry over word for word to the case of process X^λ, the λ-subprocess of X. We give, for example, the λ-variant of Theorem 4.3.1:

Theorem 4.3.5. *The following assertions are equivalent:*
1) *$(E, \mathcal{H}_{\mathcal{P}^\lambda})$ is a Brelot $\mathcal{P}$-harmonic group;*
2) *$(\dot{E}, \mathcal{H}_{\mathcal{P}^\lambda})$ is a Bauer $\mathcal{P}$-harmonic group with the Doob convergence property;*
3) *$(E, \mathcal{H}_{\mathcal{P}^\lambda_s})$ is a Brelot $\mathcal{P}$-harmonic group;*
4) *$(\dot{E}, \mathcal{H}_{\mathcal{P}^\lambda_s})$ is a Bauer $\mathcal{P}$-harmonic group with the Doob convergence property.*

4.4 Bony's theorem on the group $\mathbb{R}^p \times \mathbb{T}^\infty$

In this section we shall attempt to eliminate the condition of quasidiagonality (D). In this way we obtain a theorem on the group $E = \mathbb{R}^p \times \mathbb{T}^\infty$ similar to Bony's theorem (Theorem 2.6.2) for the group $\mathbb{R}^p$.

Theorem 4.4.1. *Let X_A and X_B be symmetric processes on group E with diffusion matrices A and B respectively. If the following conditions are satisfied:*

$$A\theta \cdot \theta \le B\theta \cdot \theta, \quad \forall \theta \in \hat{E}, \tag{4.4.1}$$

$$\sup_n |B_n|/|A_n| < \infty, \tag{4.4.2}$$

then the implication $X_B \in \mathcal{B} \Rightarrow X_A \in \mathcal{B}$ (see Sec. 3.2) holds.

The truth of the theorem is easy to deduce from the following lemma.

Lemma 4.4.1. *Let μ^A and μ^B be the symmetric Gaussian measures on E associated with matrices A and B respectively. If the conditions of Theorem 4.4.1 are satisfied, then for some number $\alpha > 0$ the following inequality is satisfied:*

$$\mu^A \leq \alpha\mu^B. \tag{4.4.3}$$

Proof. We set $\alpha^2 = \sup_n |B_n|/|A_n|$, and let μ_n^A and μ_n^B be the symmetric Gaussian measures on finite-dimensional group E_n associated with matrices A_n and B_n respectively (see (4.2.4)). From conditions (4.4.1) and (4.4.2) it follows that for every $n \geq 1$ the inequality

$$\mu_n^A \leq \alpha\mu_n^B \tag{4.4.4}$$

is satisfied. Sequences of measures $\{\mu_n^A\}_1^\infty$ and $\{\mu_n^B\}_1^\infty$ are projective with respect to system of mappings $\{\pi_{mn}\}_1^\infty$ (see Sec. 3.4.1), and measures μ^A and μ^B are their projective limits. Passing to the limit in inequality (4.4.4), we obtain (4.4.3). $\square$

Based on Theorem 4.4.1, we give the infinite-dimensional analogue of the second Bony theorem (see Theorem 2.6.2).

Theorem 4.4.2. *We shall consider an infinite-dimensional symmetric matrix $A = (a_{ij})_1^\infty$ and vector $b = (b_i)_1^\infty$. Let $\mathcal{L} = \sum_{i,j=1}^\infty a_{ij}\partial_i\partial_j + \sum_{i=1}^\infty b_i\partial_i$ and $\dot{\mathcal{L}} = \mathcal{L} - \partial_t$ be differential operators on groups E and $\dot{E}$ respectively, and $\mathcal{H}_{\mathcal{L}}$ and $\mathcal{H}_{\dot{\mathcal{L}}}$ the harmonic sheaves induced by these operators (see Sec. 4.1.3).*

If the following conditions are satisfied:

$$\rho_i < a_{ii} \ (\forall i \geq 1), \quad \rho_i := \sum_{i \neq j} |a_{ij}|, \tag{4.4.5}$$

$$\limsup_{i \to \infty} \rho_i/a_{ii} < 1, \tag{4.4.6}$$

$$\sum_{i=1}^\infty e^{-a_{ii}t} < \infty \ (\forall t > 0), \tag{4.4.7}$$

then the following assertions are true:
1) $(E, \mathcal{H}_{\mathcal{L}})$ *is a Bauer elliptic harmonic group;*
2) $(\dot{E}, \mathcal{H}_{\dot{\mathcal{L}}})$ *is a Bauer $\mathcal{P}$-harmonic group.*

If condition (4.4.5) is satisfied along with the conditions:

$$\sum_{i=1}^\infty \rho_i/a_{ii} < \infty; \tag{4.4.8}$$

$$\sum_{i=1}^\infty 1/a_{ii} < \infty, \tag{4.4.9}$$

then the following assertions are true:
3) $(E, \mathcal{H}_{\mathcal{L}})$ *is a Brelot harmonic group;*

4) $(\dot{E}, \mathcal{H}_{\dot{\mathcal{L}}})$ *is a Bauer $\mathcal{P}$-harmonic group in which the Doob convergence property is satisfied.*

Proof. From condition (4.4.5) it follows that for every $n \geq 1$, the matrix $A_n = (a_{ij})_1^n$ is positive definite (see [75: 7.2, Exercise 5]), therefore $\mathcal{L}$ is the elliptic and $\dot{\mathcal{L}}$ the parabolic differential operator (see Sec. 4.1.3). Let X be the space-homogeneous process associated with operator $\mathcal{L}$ and (μ_t) the Gaussian semigroup corresponding to this process:

$$\widehat{\mu}_t(\theta) = \exp\{-t(A\theta \cdot \theta - ib \cdot \theta)\}, \quad \theta \in \widehat{E}.$$

1. We consider assertions 1) and 2). It is obviously sufficient to show that assertion 2) is true. According to Theorem 4.2.4, we must show that for every $t > 0$ the measure μ_t is absolutely continuous with respect to Haar measure. We shall use Corollary 4.2.1. We choose a number ε such that the inequality $0 < \varepsilon < 1 - \rho_i/a_{ii}$ ($\forall i \geq 1$) is satisfied, and we consider the sequence of numbers $\{a_i\}_1^\infty : a_i = \varepsilon a_{ii}$; also let $B = \operatorname{diag}(a_i)_1^\infty$. For every $n \geq 1$, the matrix $A_n - B_n = (\tilde{a}_{ij})_1^n$ is constructed in the following manner:

$$\tilde{a}_{ij} = \begin{cases} a_{ij}, & i \neq j \\ (1 - \varepsilon)a_{ii}, & i = j. \end{cases}$$

By the choice of ε we conclude that the inequality

$$\tilde{a}_{ii} > \sum_{i \neq j} |\tilde{a}_{ij}|, \quad i = 1, \ldots, n,$$

is satisfied, i.e. $A_n - B_n$ is a matrix with dominating diagonal. According to [75: 7.2, Exercise 5] from this property it follows that the matrix $A_n - B_n$ is positive definite. Consequently the condition

$$\sum_{k=1}^\infty a_k \theta_k^2 \leq A\theta \cdot \theta, \quad \theta \in \widehat{E} = \mathbb{R}^p \times \mathbb{Z}^{(\infty)}$$

is satisfied. According to condition (4.4.7),

$$\sum_{k=1}^\infty e^{-a_k t} = \sum_{k=1}^\infty e^{-a_{kk}\varepsilon t} < \infty \quad (\forall t > 0).$$

Now applying Corollary 4.2.1, we conclude that for every $t > 0$ the measure μ_t is absolutely continuous with respect to Haar measure.

2. We consider assertions 3) and 4). It is obvious that it is sufficient to show the truth of assertion 4). According to Theorem 4.2.5, we must show that the function

$$\dot{r}(\dot{x}) = \begin{cases} \mu_t(x), & t > 0 \\ 0, & t \leq 0 \end{cases}$$

is continuous on $\dot{E}\backslash\{\dot{0}\}$. By relation (4.2.2) it is sufficient to establish this property assuming that measure μ_t is symmetric, i.e. $b = 0$ and thus $X := X_A$ is a symmetric process. For a symmetric process the continuity of function $\dot{r}(\dot{x})$ is equivalent to the fact that this process is in class $\mathcal{B}$ (see Remark 4.3.1). We shall use Theorem 4.4.1. We choose an $\varepsilon > 0$ and set

$$\Delta_i := \left(1 + \frac{\rho_i + \varepsilon}{a_{ii}}\right), \quad \beta_i := \Delta_i a_{ii}, \quad i = 1, 2, \ldots$$

and consider the diagonal matrix $B = \operatorname{diag}(\beta_i)_1^\infty$. Let X_B be the symmetric process associated with matrix B. From (4.4.9) it follows that $\sum_{i=1}^\infty 1/\beta_i < \infty$, therefore by Theorem 4.3.3 the process X_B is in class $\mathcal{B}$. We shall show that the conditions of Theorem 4.4.1 are satisfied. For every $n \geq 1$ the matrix $B_n - A_n = (\tilde{a}_{ij})_1^n$ is constructed in the following manner:

$$\tilde{a}_{ij} = \begin{cases} -a_{ij}, & i \neq j \\ \rho_i + \varepsilon, & i = j. \end{cases}$$

From this it follows that the inequality

$$\tilde{a}_{ii} > \sum_{i \neq j} |\tilde{a}_{ij}|, \quad i = 1, \ldots, n,$$

is satisfied, i.e. the matrix $B_n - A_n$ has a dominating principal diagonal. Consequently matrix $B_n - A_n$ is positive definite ($\forall n \geq 1$), i.e. condition (4.4.1) is satisfied. We shall show that condition (4.4.2) is also satisfied. From condition (4.4.5) it follows (see [75: 7.2, Exercise 4]) that the inequality

$$|A_n| \geq \prod_{i=1}^n (a_{ii} - \rho_i)$$

is satisfied, with the help of which it is now easy to verify condition (4.4.2):

$$\sup_n |B_n|/|A_n| \leq \sup_n \prod_{i=1}^n \Delta_i a_{ii} / \prod_{i=1}^n (a_{ii} - \rho_i) \leq$$

$$\leq \prod_{i=1}^\infty \left(1 + \frac{\rho_i}{a_{ii}} + \frac{\varepsilon}{a_{ii}}\right) \prod_{i=1}^\infty \left(1 + \frac{\rho_i}{a_{ii}} \cdot \left[1 - \frac{\rho_i}{a_{ii}}\right]^{-1}\right) < \infty.$$

Thus the conditions of Theorem 4.4.1 are satisfied and consequently $X_A \in \mathcal{B}$. $\square$

Remark 4.4.1. Conditions (4.4.7) and (4.4.9) are exact in the following sense. Let $A = \operatorname{diag}(a_k)_1^\infty$ and $b = 0$, i.e. $\mathcal{L} = \sum_{k=1}^\infty a_k \partial_k^2$. We shall show that the following assertions are true:[3]

1. $a_k \preceq \ln k \Rightarrow (E, \mathcal{H}_\mathcal{L})$ is not a Bauer harmonic group;

---

3    $a_k \prec b_k \Leftrightarrow \frac{a_k}{b_k} = o(1)$; $a_k \asymp b_k \Leftrightarrow \frac{a_k}{b_k} = O(1)$ and $\frac{b_k}{a_k} = O(1)$.

2. $\ln k \prec a_k \preceq k \Rightarrow (E, \mathcal{H}_{\mathscr{L}})$ is a Bauer elliptic harmonic group and is not a Brelot harmonic group;

3. $k \prec a_k \Rightarrow (E, \mathcal{H}_{\mathscr{L}})$ is a Brelot harmonic group.

1. According to Theorem 4.3.2, measure μ_t is not absolutely continuous with respect to Haar measure for all $t \in]0, \delta^{-1}]$, where $\delta = \sup_k a_k / \ln k$. By Proposition 4.2.1, from this it follows that the measure $r = \int_0^\infty \mu_t \, dt$ is not absolutely continuous with respect to Haar measure. Therefore, by Theorem 4.2.1, $(E, \mathcal{H}_{\mathscr{L}})$ is not a harmonic group.

2. By Theorem 4.3.2, measure μ_t is absolutely continuous with respect to Haar measure for all $t > 0$, and consequently measure r is also absolutely continuous. By Theorem 4.2.2, $(E, \mathcal{H}_{\mathscr{L}})$ is a Bauer elliptic harmonic group. According to Remark 4.3.1 and Theorem 4.3.3, $(E, \mathcal{H}_{\mathscr{L}})$ is not a Brelot harmonic group.

3. According to Remark 4.3.1 and Theorem 4.3.3, $(E, \mathcal{H}_{\mathscr{L}})$ is a Brelot harmonic group. $\qquad\square$

Remark 4.4.2. Theorem 4.4.2 remains true if instead of differential operators $\mathscr{L}$ and $\dot{\mathscr{L}}$, we consider differential operators $\mathscr{L}^\lambda = \mathscr{L} - \lambda I$ and $\dot{\mathscr{L}}^\lambda = \dot{\mathscr{L}} - \lambda I$, and $\mathcal{H}_{\mathscr{L}^\lambda}$ and $\mathcal{H}_{\dot{\mathscr{L}}^\lambda}$ are sheaves of weak solutions of equations $\mathscr{L}u - \lambda u = 0$ and $\dot{\mathscr{L}}u - \lambda u = 0$ respectively.

We finish this section showing that there is a close connection between the topological structure of a group and the properties of harmonic functions on it. The existence of this relationship was conjectured by Bliedtner in 1969 in his paper *On the analytic structure of harmonic groups* [37].

Theorem 4.4.3. *Let E be a locally compact abelian group. The following statements are equivalent:*

1. *E is locally Euclidean.*

2. *The concepts of elliptic harmonic group and Brelot harmonic group coincide on E.*

Proof. 1. $\Rightarrow$ 2. follows from Bony's theorem. 2. $\Rightarrow$ 1.: according to [46: p. 31] the group E is locally connected. Let E_0 be the component of its neutral element. By Dixmier's theorem [63: Theorem B] we have $E_0 \cong \mathbb{R}^p \times \mathbb{T}^m$ where $p < \infty$, $m \leq \infty$. Now it remains to apply Remark 4.4.1. $\qquad\square$

Chapter 5
Elliptic equations on a group

In the previous chapter we investigated continuity and convergence properties of the sheaf $\mathscr{H}_{\mathscr{L}\lambda}$ of λ-harmonic functions of a space-homogeneous process X with almost surely continuous trajectories on group $E = \mathbb{R}^p \times \mathbb{T}^\infty$. Every function $u \in \mathscr{H}_{\mathscr{L}\lambda}(V)$, as a generalized function, satisfies the equation $\mathscr{L}u - \lambda u = 0$ on V, where $\mathscr{L} = \sum_{i,j=1}^\infty a_{ij}\partial_i\partial_j + \sum_{i=1}^\infty b_i\partial_i$ is the infinite-dimensional elliptic differential operator associated with X. Here we concentrate our attention on the properties of differentiability of generalized solutions of this equation. Our investigation is based on some *apriori* inequalities which are given in Secs. 5.1 and 5.4. In Sec. 5.2 we show that every function $u \in \mathscr{H}_{\mathscr{L}\lambda}(V)$ has generalized derivatives in L_p $(p > 1)$ of every order. We note that, in distinction from the finite-dimensional case, the function u in this connection need not even be continuous. According to the results of Sec. 5.3, continuous differentiability of every weak solution (more precisely, the existence of a continuously differentiable regularization) of equation $\mathscr{L}u - \lambda u = 0$ is equivalent to the process X belonging to class $\mathscr{B}$ (see Sec. 3.2).

For simplicity we shall assume that $E = \mathbb{T}^\infty$ and X is a symmetric process with diagonal diffusion matrix $A = \mathrm{diag}(a_k)_1^\infty$ (for the general case see [14, 16]). In correspondence with this assumption, $\mathscr{L} = \sum_{k=1}^\infty a_k\partial_k^2$.

5.1 Admissible distributions and multipliers

5.1.1

Let (μ_t) be the Gaussian semigroup of measures associated with operator $\mathscr{L} = \sum_{k=1}^\infty a_k\partial_k^2$:

$$\widehat{\mu}_t(\theta) = \exp(-tA\theta \cdot \theta), \quad \theta \in \mathbb{Z}^{(\infty)}, \quad A = \mathrm{diag}(a_k)_1^\infty$$

and (q_t) the Cauchy semigroup corresponding to semigroup (μ_t) (Sec. 4.2.4):

$$q_t = \int_0^\infty \mu_s h_t(s)\, ds, \quad h_t(s) = (4\pi)^{-1/2} t \cdot s^{-3/2} \exp\left(-\frac{t^2}{4s}\right).$$

We denote by (P_t) and (Q_t) the semigroups of contractions on $L_p = L_p(T^\infty; dx)$ induced by semigroups of measures (μ_t) and (q_t) respectively

$$P_t f = f * \mu_t, \quad Q_t f = f * q_t, \quad f \in L_p$$

and let $\mathscr{L}_p$ and B_p be L_p-generators of these semigroups. It is easy to show that the following properties are satisfied on the set $\mathscr{D}$ of cylindrical infinitely differentiable functions:

1) $\mathscr{L}_p|_{\mathscr{D}} = \mathscr{L}$,
2) $B_p : \mathscr{D} \to \mathscr{D}$,
3) if $B := B_p|_{\mathscr{D}}$, then $B^2 = -\mathscr{L}$.

Following Meyer [79: p. 164], we introduce into consideration the field operator

$$\Gamma(f, g) = A\nabla f \cdot \nabla g, \quad f, g \in \mathscr{D}, \quad \nabla := (\partial_k)_1^\infty$$

and the Littlewood–Paley functions: $f \in \mathscr{D}$,

$$G_\to f = \left\{ \int_0^\infty (\partial_t Q_t f)^2 t \, dt \right\}^{1/2},$$

$$Gf = \left\{ \int_0^\infty [\Gamma(Q_t f, Q_t f) + (\partial_t Q_t f)^2] t \, dt \right\}^{1/2}.$$

It is obvious that $G_\to f \leq Gf$. The proof of the following theorem is obtained from the results of [79: pp. 165–168].

Theorem 5.1.1. *For $1 < p < \infty$, the following inequality holds:*

$$\|Gf\|_{L_p} \leq c_p \|f\|_{L_p}. \tag{5.1.1}$$

5.1.2

Let $f = (f_i)_1^\infty$ be a vector function on T^∞. We set

$$|f| = \left(\sum_{k=1}^\infty |f_k|^2 \right)^{1/2}, \quad \vec{L}_p = \{ f = (f_k)_1^\infty : |f| \in L_p \}.$$

For a vector function $f = (f_k)_1^\infty$ for which all components f_k lie in $\mathscr{D}$, we define

$$G_\to f := \left\{ \int_0^\infty |\partial_t Q_t f|^2 t \, dt \right\}^{1/2}.$$

Theorem 5.1.2. *For $1 < p < \infty$, the following inequality holds:*

$$\|G_\to f\|_{L_p} \leq c_p \|f\|_{\vec{L}_p}. \tag{5.1.2}$$

Proof. We shall carry out the arguments assuming that the vector function f has a finite number of non-zero components (then monotonic limit passage can be applied). We shall denote the constants depending on p by c_p^k $(k = 1, 2, \ldots)$.

Let $s \in [0, 1]$ and $r(s) = (r_1(s), r_2(s), \ldots)$ be a vector consisting of Rademacher functions [101: p. 124]. Applying inequality (5.1.1) to scalar function $r \cdot f$, we obtain

$$\|G_\to (r \cdot f)\|_{L_p}^p \leq c_p^1 \|r \cdot f\|_{L_p}^p.$$

Integrating this inequality over the interval $[0, 1]$, making the obvious transformations and applying the Khinchin inequality [101: p. 124 (44)], we obtain

$$\int_0^1 ds\, \|G_\to (r \cdot f)\|_{L_p}^p \leq c_p^1 \int_0^1 ds\, \|r \cdot f\|_{L_p}^p =$$

$$= c_p^1 \int_0^1 ds \int_{\mathbb{T}^\infty} dx |r \cdot f|^p = c_p^1 \int_{\mathbb{T}^\infty} dx \int_0^1 ds |r \cdot f|^p \leq$$

$$\leq c_p^2 \int_{\mathbb{T}^\infty} dx |f|^p = c_p^2 \|f\|_{L_p}^p.$$

Thus, we obtain the inequality

$$\int_0^1 ds\, \|G_\to (r \cdot f)\|_{L_p}^p \leq c_p^2 \|f\|_{L_p}^p. \tag{5.1.3}$$

Now we estimate the integral on the left-hand side of inequality (5.1.3) from below. We have

$$\int_0^1 ds\, \|G_\to (r \cdot f)\|_{L_p}^p = \int_0^1 ds \int_{\mathbb{T}^\infty} dx (G_\to (r \cdot f))^p = \tag{5.1.4}$$

$$= \int_{\mathbb{T}^\infty} dx \int_0^1 ds (G_\to (r \cdot f))^p \geq \int_{\mathbb{T}^\infty} dx \left(\int_0^1 ds G_\to (r \cdot f) \right)^p.$$

For convenience we temporarily introduce the following notation:

$$\mu(dt) = t\, dt, \quad g = \partial_t Q_t f,$$

$$\varphi = |\partial_t Q_t (r \cdot f)| = |r \cdot \partial_t Q_t f| = |r \cdot g|,$$

$L_2(\mu)$ is the space of functions on $[0, \infty[$ that are square-integrable with respect to measure μ, $\langle x, y \rangle_\mu$ is the scalar product in $L_2(\mu)$, and $\|x\|_\mu^2 = \langle x, x \rangle_\mu$.

Assuming that the function h runs over the unit sphere in $L_2(\mu)$, we carry out the further estimates

$$\int_0^1 ds G_\to (r \cdot f) = \int_0^1 ds \left(\int_0^\infty (\partial_t Q_t (r \cdot f))^2 t\, dt \right)^{1/2} =$$

$$= \int_0^1 ds \, \|\varphi\|_\mu = \int_0^1 ds \sup_h |\langle \varphi, h \rangle_\mu| \geq$$

$$\geq \sup_h \left| \int_0^1 ds \int_0^\infty \varphi \cdot h \, d\mu \right| = \sup_h \left| \int_0^\infty h \, d\mu \int_0^1 \varphi \, ds \right| =$$

$$= \sup_h \left| \int_0^\infty h \, d\mu \int_0^1 |r \cdot g| \, ds \right| \geq c_1 \sup_h \left| \int_0^\infty h |g| \, d\mu \right| =$$

$$= c_1 \, \| \, |g| \, \|_\mu = c_1 G_\to f.$$

Thus we arrive at the inequality

$$\int_0^1 ds G_\to (r \cdot f) \geq c_1 G_\to f.$$

With the help of this inequality we continue (5.1.4)

$$\int_0^1 ds \, \|G_\to (r \cdot f)\|_{L_p}^p \geq c_p^3 \int_{\mathbb{T}^\infty} dx (G_\to f)^p = c_p^3 \|G_\to f\|_{L_p}^p. \tag{5.1.5}$$

Comparing (5.1.3) and (5.1.5), we obtain the required inequality. $\square$

Theorem 5.1.3. *If* $1 < p < \infty$ *and the vector-function* $f = (f_i)_1^\infty$ *is such that* $\langle f_i, 1 \rangle = 0$ *(*$\forall i \geq 1$*), then the following inequality holds:*

$$\|f\|_{\vec{L}_p} \leq c_p \|G_\to f\|_{L_p}. \tag{5.1.6}$$

Proof. As in the proof of Theorem 5.1.2, we shall carry out arguments assuming that f is a finite-dimensional vector. Let the number q be such that $1/p + 1/q = 1$, and $g = (g_i)$ is a finite-dimensional vector whose components g_i lie in $\mathcal{D}$. Applying Parseval's equality and the orthogonality of the components of vector f to the unit, as well as setting $\Psi(\theta) = A\theta \cdot \theta$, we get

$$\langle f, g \rangle = \langle \hat{f}, \hat{g} \rangle = \int_0^\infty t \, dt \left\langle -\sqrt{\Psi} e^{-t\sqrt{\Psi}} \hat{f}, -\sqrt{\Psi} e^{-t\sqrt{\Psi}} \hat{g} \right\rangle =$$

$$= \int_0^\infty \langle \widehat{\partial_t Q_t f}, \widehat{\partial_t Q_t g} \rangle t \, dt = \int_{\mathbb{T}^\infty} dx \int_0^\infty (\partial_t Q_t f)(\partial_t Q_t g) t \, dt.$$

Therefore, considering (5.1.2), we obtain

$$|\langle f, g \rangle| \leq \int_{\mathbb{T}^\infty} dx \left(\int_0^\infty (\partial_t Q_t f)^2 t \, dt \right)^{1/2} \left(\int_0^\infty (\partial_t Q_t g)^2 t \, dt \right)^{1/2} =$$

$$= \langle G_\to f, G_\to g \rangle \leq \|G_\to f\|_{L_p} \|G_\to g\|_{L_q} \leq c_q \|G_\to f\|_{L_p} \|g\|_{\vec{L}_q}.$$

Thus, we have obtained the inequality

$$|\langle f, g \rangle| \leq c_q \|G_\to f\|_{L_p} \|g\|_{\vec{L}_q}.$$

Since the set of vector-functions g with components from $\mathcal{D}$ is dense in $\vec{L}_q$, the required inequality is established. $\square$

5.1.3

We introduce the concept of convergence on linear space $\mathscr{D}$. We shall say that a sequence $\{\varphi_n\}_1^\infty$ converges to zero in $\mathscr{D}$ if all functions φ_n depend on a finite number of fixed variables $x_1, \ldots, x_m$ and for any multi-index $\alpha = (\alpha_1, \ldots, \alpha_m)$ we have

$$\lim_{n\to\infty} \sup_{x\in\mathbb{T}^\infty} |\partial^\alpha \varphi_n(x)| = 0, \quad \partial^\alpha = \partial_1^{\alpha_1} \ldots \partial_m^{\alpha_m}.$$

The set of all linear functionals on $\mathscr{D}$ that are continuous with respect to this convergence shall be denoted by $\mathscr{D}^*$ and shall be called the space of distributions (generalized functions) on $\mathbb{T}^\infty$. We introduce two examples of distributions:
1) for function $f \in L_1$, we define a linear functional Λ_f by the relation

$$\Lambda_f(\varphi) = \langle f, \varphi \rangle \quad (\varphi \in \mathscr{D});$$

2) for measure μ we define a linear functional Λ_μ:

$$\Lambda_\mu(\varphi) = \langle \mu, \varphi \rangle \quad (\varphi \in \mathscr{D}).$$

In the following, instead of notation Λ_f and Λ_μ we shall use the notation f and μ; this leads to no ambiguity.

Some analysis operations are introduced on space $\mathscr{D}^*$ in the standard manner: differentiation, convolution, Fourier transformation.

It is possible to associate a linear continuous operator $L : \mathscr{D} \to \mathscr{D}$ with each distribution Λ:

$$L\varphi(x) = \Lambda * \varphi(x) \quad (\varphi \in \mathscr{D}).$$

In particular, if $\Lambda = \partial_k \varepsilon_0$, ε_0 the Dirac measure at zero, then the corresponding operator, as it is not difficult to see, coincides with the differentiation operator ∂_k with respect to variable x_k.

The vector $\vec{\Lambda} = (\Lambda_k)_1^\infty$, $\Lambda_k \in \mathscr{D}^*$ ($\forall k \geq 1$) is called admissible if for any function $\varphi \in \mathscr{D}$ the following inequality is satisfied:

$$|\vec{\Lambda}(\varphi)|^2 = \sum_{k=1}^\infty |\Lambda_k(\varphi)|^2 \leq \Gamma(\varphi, \varphi)(0) + |B\varphi(0)|^2. \tag{5.1.7}$$

We associate a linear operator $\vec{L} : \mathscr{D} \to \vec{L}_p$ with each admissible vector $\vec{\Lambda}$ by the formula

$$\vec{L}\varphi := \vec{\Lambda} * \varphi = (\Lambda_k * \varphi)_1^\infty.$$

We omit the standard proof of the following Proposition 5.1.1:

Proposition 5.1.1. *Let $\vec{\Lambda}$ be an admissible vector and $\vec{L}$ the linear operator corresponding to $\vec{\Lambda}$. The following properties hold:*
1) $Q_t\vec{L}\varphi = \vec{L}Q_t\varphi$ $(\forall\varphi \in \mathcal{D})$;
2) $|\vec{L}\varphi|^2 \leq \Gamma(\varphi, \varphi) + |B\varphi|^2$ $(\forall\varphi \in \mathcal{D})$;
3) *the components of vector $\vec{L}\varphi$ are orthogonal to the unit.*

Theorem 5.1.4. *Let $\vec{\Lambda}$ be an admissible vector and $\vec{L}$ the linear operator corresponding to $\vec{\Lambda}$. For any function $\varphi \in \mathcal{D}$, the following inequality holds:*

$$\|\vec{L}\varphi\|_{\vec{L}_p} \leq c_p\|B\varphi\|_{L_p} \ (1 < p < \infty).$$

Proof. Differentiating both sides of equality 1) of Proposition 5.1.1 with respect to t, we get

$$\partial_t Q_t\vec{L}\varphi = \partial_t\vec{L}Q_t\varphi = \vec{L}\partial_t Q_t\varphi = \vec{L}Q_t B\varphi,$$

or, denoting $r = \vec{L}\varphi$, $g = B\varphi$,

$$\partial_t Q_t r = \vec{L}Q_t g.$$

We scalar-square both sides of this equality and apply property 2) of Proposition 5.1.1:

$$(\partial_t Q_t r)^2 = (\vec{L}Q_t g)^2 \leq \Gamma(Q_t g, Q_t g) + (BQ_t g)^2 = \Gamma(Q_t g, Q_t g) + (\partial_t Q_t g)^2.$$

We integrate the obtained inequality on the interval $[0, \infty[$ with respect to measure $t\,dt$:

$$\int_0^\infty (\partial_t Q_t r)^2 t\,dt \leq \int_0^\infty [\Gamma(Q_t g, Q_t g) + (\partial_t Q_t g)^2]t\,dt.$$

According to the notation of Secs. 5.1.1-5.1.2, the function $(Gg)^2$ is on the right-hand side, and on the left, $(G_{\rightarrow}r)^2$. Thus $G_{\rightarrow}r \leq Gg$, from which we obtain the inequality

$$\|G_{\rightarrow}r\|_{L_p} \leq \|Gg\|_{L_p}.$$

According to Theorem 5.4.1,

$$\|Gg\|_{L_p} \leq c_p\|g\|_{L_p}.$$

By Proposition 5.1.1 3), the components of vector r are orthogonal to the unit, therefore according to Theorem 5.4.3,

$$\|r\|_{\vec{L}_p} \leq c_p\|G_{\rightarrow}r\|_{L_p}.$$

Summing up the whole account, we write the chain of inequalities

$$\|r\|_{\vec{L}_p} \leq c_p\|G_{\rightarrow}r\|_{L_p} \leq c_p\|Gg\|_{L_p} \leq c_p^1\|g\|_{L_p},$$

from which, considering that $r = \vec{L}\varphi$ and $g = B\varphi$, we obtain the required inequality. $\qquad\qquad\square$

A simple but important example of an admissible vector is the vector

$$\vec{\Lambda} = (\Lambda_k)_1^\infty : \Lambda_k = \sqrt{a_k}\partial_k\varepsilon_0, \quad k = 1, 2, \ldots$$

We shall denote the corresponding operator $\vec{L}$ by ∇_A, where $A = \mathrm{diag}(a_k)_1^\infty$.

Theorem 5.1.5. *For any function $\varphi \in \mathcal{D}$, and $1 < p < \infty$, the following inequality holds:*

$$c_p^{-1}\|B\varphi\|_{L_p} \;\leq\; \|\nabla_A\varphi\|_{\vec{L}_p} \;\leq\; c_p\|B\varphi\|_{L_p}.$$

Proof. The inequality

$$\|\nabla_A\varphi\|_{\vec{L}_p} \leq c_p\|B\varphi\|_{L_p}$$

holds by Theorem 5.1.4. We proceed with the proof of the reverse inequality. Applying Parseval's equality we obtain

$$\langle B\varphi, Bh \rangle = \int_{\mathbb{T}^\infty} \nabla_A\varphi \cdot \nabla_A h \, dx.$$

Let $f \in \mathcal{D}$ and (V_λ) be the resolvent of semigroup (Q_t). We set

$$h = -V_\lambda f, \quad g = Bh = f - \lambda V_\lambda f$$

and we write the following chain of inequalities: $(1/p + 1/q = 1)$

$$|\langle B\varphi, g \rangle| \;=\; |\langle B\varphi, Bh \rangle| \;=\; \left|\int_{T^\infty} \nabla_A\varphi \cdot \nabla_A h \, dx\right| \;\leq$$

$$\leq \int_{T^\infty} |\nabla_A\varphi||\nabla_A h|dx \;\leq\; \|\nabla_A\varphi\|_{\vec{L}_p}\|\nabla_A h\|_{\vec{L}_q} \;\leq$$

$$\leq c_q\|\nabla_A\varphi\|_{\vec{L}_p}\|Bh\|_{L_q} = c_q\|\nabla_A\varphi\|_{\vec{L}_p}\|g\|_{L_q}.$$

Further, we have

$$\lim_{t\to\infty} Q_t f(x) = \lim_{t\to\infty} \int_{\mathbb{Z}^{(\infty)}} e^{-t\sqrt{\Psi(\theta)}} \widehat{f}(\theta)e^{i\theta\cdot x}d\theta = \widehat{f}(0) = \langle f, 1 \rangle.$$

Therefore

$$\lim_{\lambda\to 0} \lambda V_\lambda f(x) = \lim_{\lambda\to 0} \int_0^\infty e^{-t}Q_{t/\lambda}f(x)\,dt = \langle f, 1 \rangle.$$

Consequently for $\lambda \to 0$ we have $g \to f - \langle f, 1 \rangle$ and we arrive at the inequality

$$|\langle B\varphi, f - \langle f, 1 \rangle \rangle| \leq c_q\|\nabla_A\varphi\|_{\vec{L}_p}\|f - \langle f, 1 \rangle\|_{L_q}.$$

Now we note that $\langle B\varphi, 1\rangle = 0$ and $\|f - \langle f, 1\rangle\|_{L_q} \le 2\|f\|_{L_q}$, and therefore the following inequality is established:

$$|\langle B\varphi, f\rangle| \le c_q^1 \|\nabla_A \varphi\|_{\vec{L}_p} \|f\|_{L_q} \quad (\forall f \in \mathscr{D}),$$

which clearly finishes the proof of the theorem. $\qquad\qquad\square$

5.1.4

Let a vector function $m(\theta) = (m_i(\theta))_1^\infty$ be defined on $\mathbb{Z}^{(\infty)}$, where its modulus is bounded. We define the linear transformation $T = (T_i)_1^\infty$ on $\mathscr{D}$ in the following manner:

$$\widehat{Tf}(\theta) = m(\theta)\widehat{f}(\theta), \quad \theta \in \mathbb{Z}^{(\infty)}. \tag{5.1.8}$$

We shall call m an $(L_p, \vec{L}_p)$-multiplier, and T a multiplier operator from L_p to $\vec{L}_p$, if for any function $f \in \mathscr{D}$, the following inequality is satisfied:

$$\|Tf\|_{\vec{L}_p} \le c_p \| f \|_{L_p} \quad (1 < p < \infty). \tag{5.1.9}$$

Remark 5.1.1. Equality (5.1.8) correctly defines the vector function $Tf \in \vec{L}_2$ for $f \in \mathscr{D}$, where the components of this vector function lie in $\mathscr{D}$. From inequality (5.1.9) and from the density of $\mathscr{D}$ in L_p ($1 \le p < \infty$) it follows that operator T can be continued by continuity up to a bounded operator from L_p to $\vec{L}_p$.

Theorem 5.1.6. *Let* $\vec{\Lambda} = (\Lambda_k)_1^\infty$ *be an admissible vector of distributions and* $\vec{\lambda}(\theta) = (\lambda_k(\theta))_1^\infty$ *its Fourier transform. Then the vector function on* $\mathbb{Z}^{(\infty)}$:

$$m(\theta) = \begin{cases} \vec{\lambda}(\theta)/\sqrt{\Psi(\theta)}, & \theta \ne 0 \\ 0, & \theta = 0 \end{cases} \tag{5.1.10}$$

is an $(L_p, \vec{L}_p)$-*multiplier for* $1 < p < \infty$.

Proof. Since $\vec{\Lambda}$ is an admissible vector and $\vec{\lambda}(\theta) = \vec{\Lambda}(e^{i\theta\cdot})$, then using inequality (5.1.7), we get

$$|\vec{\lambda}(\theta)|^2 = |\vec{\Lambda}(e^{i\theta\cdot})|^2 \le \Gamma(e^{i\theta\cdot}, e^{i\theta\cdot})(0) + |Be^{i\theta\cdot}(0)|^2 = 2\Psi(\theta).$$

From this it follows that the vector-function m is bounded.

Let $f \in \mathscr{D}$. We consider the function g on $\mathbb{T}^\infty$, which we define by the Fourier transform:

$$\widehat{g}(\theta) = \begin{cases} \widehat{f}(\theta)/\sqrt{\Psi(\theta)}, & \theta \ne 0 \\ 0, & \theta = 0. \end{cases}$$

We set $\mathbb{Z}^{(n)} = \{\theta = (\theta_i)_1^\infty \in \mathbb{Z}^{(\infty)} : \theta_i = 0, \forall i > n\}$; then it is obvious that:

$$\mathbb{Z}^{(1)} \subset \mathbb{Z}^{(2)} \subset \dots, \quad \cup_{n=1}^\infty \mathbb{Z}^{(n)} = \mathbb{Z}^{(\infty)};$$

in other words the group $\mathbb{Z}^{(\infty)}$ is the inductive limit of the sequence of its subgroups $\{\mathbb{Z}^{(n)}\}_1^\infty$. Now noting that the support of function $\widehat{f}$ is contained in $\mathbb{Z}^{(n)}$ for some $n \geq 1$, we carry out an estimate of function $\widehat{g}(\theta)$ for $\theta \neq 0$:

$$\begin{aligned}
|\widehat{g}(\theta)| &= |\widehat{f}(\theta)/\sqrt{\Psi(\theta)}| = |\widehat{f}(\theta)1_{\mathbb{Z}^{(n)}}(\theta)/\sqrt{\Psi(\theta)}| = \\
&= |\widehat{f}(\theta)|1_{\mathbb{Z}^{(n)}}(\theta)/\sqrt{\Psi(\theta)} \leq |\widehat{f}(\theta)| \sup_{k \leq n}(1/\sqrt{a_k}).
\end{aligned}$$

From this estimate and from the fact that $\widehat{f} \in L_1(\mathbb{Z}^{(\infty)})$, it also follows that $\widehat{g} \in L_1(\mathbb{Z}^{(\infty)})$, and therefore the function g is defined uniquely and furthermore is continuous. We note also that g is a cylindrical function depending on the same number of variables as f. We shall show that $g \in \mathcal{D}$. In fact, for $k \geq 0$ and $\theta \neq 0$, we have

$$|\Psi^k \cdot \widehat{g}| \leq \sup_{k \leq n}(1/\sqrt{a_k})|\Psi^k \widehat{f}|.$$

Now it remains to note that a cylindrical function φ is in $\mathcal{D}$ if and only if for any $k \geq 0$ the function $\Psi^k \widehat{\varphi}$ is in $L_1(\mathbb{Z}^{(\infty)})$.

From the above, it follows that it is possible to apply to function g an operator $\vec{L}$ generated by vector $\vec{\Lambda}$. Passing to the Fourier transform, it is easy to verify the equality

$$Tf = \vec{L}g,$$

where the operator T is defined by formulas (5.1.8) and (5.1.10). Applying Theorem 5.1.4, for $1 < p < \infty$ we obtain

$$||Tf||_{\vec{L}_p} = ||\vec{L}g||_{\vec{L}_p} \leq c_p||Bg||_{L_p}.$$

However,

$$\begin{aligned}
Bg(x) &= \int_{\mathbb{Z}^{(\infty)}} e^{i\theta \cdot x}\widehat{Bg}(\theta)d\theta = \int_{\mathbb{Z}^{(\infty)}\setminus\{0\}} e^{i\theta \cdot x}\sqrt{\Psi(\theta)}\frac{\widehat{f}(\theta)}{\sqrt{\Psi(\theta)}}d\theta = \\
&= \int_{\mathbb{Z}^{(\infty)}\setminus\{0\}} e^{i\theta \cdot x}\widehat{f}(\theta)d\theta = \int_{\mathbb{Z}^{(\infty)}} e^{i\theta \cdot x}\widehat{f}(\theta)d\theta - \widehat{f}(0) = \\
&= f(x) - \langle f, 1\rangle.
\end{aligned}$$

Therefore,

$$||Tf||_{\vec{L}_p} \leq c_p||f - \langle f, 1\rangle||_{L_p} \leq c_p^1||f||_{L_p},$$

consequently m is an $(L_p, \vec{L}_p)$-multiplier for all $1 < p < \infty$. $\qquad \square$

Remark 5.1.2. From the structure of the multiplier m associated with admissible vector of distributions $\vec{\Lambda}$, it follows that

$$\langle T_k f, 1 \rangle = 0 \quad (\forall k \geq 1).$$

5.1.5

For the function $f \in \mathcal{D}$ we set

$$\Lambda_k f := -\sqrt{a_k}\partial_k f(0) \quad (k = 1, 2, \ldots)$$

and let $\vec{\Lambda} = (\Lambda_k)_1^\infty$. Since

$$\sum_{k=1}^\infty (\Lambda_k f)^2 = \sum_{k=1}^\infty a_k (\partial_k f(0))^2 = \Gamma(f, f)(0),$$

then vector $\vec{\Lambda}$ is admissible (see Sec. 5.1.7)). We calculate the Fourier transform $\lambda_k(\theta)$ of distribution Λ_k:

$$\lambda_k(\theta) = \Lambda_k(e^{i\theta \cdot}) = -i\sqrt{a_k}\theta_k, \quad \theta = (\theta_i)_1^\infty.$$

The multiplier m_k associated with distribution Λ_k has the form

$$m_k(\theta) = \begin{cases} -i\sqrt{a_k}\theta_k/\sqrt{\Psi(\theta)}, & \theta \neq 0 \\ 0, & \theta = 0. \end{cases}$$

We shall denote the operator corresponding to multiplier m_k by R_k and call it the k-th Riesz operator on $\mathbb{T}^\infty$, and operator $\vec{R} = (R_i)_1^\infty$ is the Riesz vector operator on $\mathbb{T}^\infty$.

On the basis of Theorem 5.1.6, we have

Theorem 5.1.7. *For $1 < p < \infty$, the following inequality holds:*

$$\|\vec{R}f\|_{\vec{L}_p} \leq c_p \|f\|_{L_p}. \tag{5.1.11}$$

Remark 5.1.3. It is easy to see that for any $k \geq 1$ and for any $n \geq 1$, the operator R_k transforms the set of cylindrical functions depending on the first n variables into itself. Therefore the following operator $R_{kn} : L_p(\mathbb{T}^n) \to L_p(\mathbb{T}^n)$ is correctly defined by

$$(R_{kn}\varphi) \cdot \pi_{n\infty} := R_k(\varphi \cdot \pi_{n\infty}).$$

Now we note that for $k > n$ we have $R_{kn} = 0$. If we then set $a_1 = a_2 = \ldots = a_n = 1$, then the operators $R_{1n}, \ldots, R_{nn}$ coincide with the classical Riesz operators on $\mathbb{T}^n$ (see [102: Chapter 7]). Setting $R_{(n)} = (R_{1n}, \ldots, R_{nn})$, in correspondence with Theorem 5.1.7, we write the inequality

$$\|R_{(n)}f\|_{\vec{L}_p(T^n)} \leq c_p \|f\|_{L_p(\mathbb{T}^n)}$$

in which the constant c_p does not depend on the dimension n.

We note that the analogous inequality (but for group $\mathbb{R}^n$) was established in 1983 by Stein [103] (see also [12], [15]).

5.1.6

In conclusion of this section we give a theorem which shows the importance of Theorems 5.1.5 and 5.1.7.

We denote by W_p^k ($k = 1, 2, \ldots$) the linear space of functions of L_p whose generalized derivatives up to the k-th order inclusive are in L_p.

We shall say that a function $u \in L_p$ is a weak solution of the equation

$$\mathscr{L}u - \lambda u = f, \quad f \in L_p, \tag{5.1.12}$$

if for any function $\varphi \in \mathscr{D}$ the following relation is satisfied:

$$\langle u, \mathscr{L}\varphi - \lambda\varphi \rangle = \langle f, \varphi \rangle. \tag{5.1.13}$$

Theorem 5.1.8. *Every weak solution u of equation (5.1.12) with right-hand side $f \in W_p^k$ belongs to W_p^{k+2}, and $u \in dom(\mathscr{L}_p)$ (see Sec. 5.1.1),*

$$\mathscr{L}_p u = \sum_{k=1}^{\infty} a_k \partial_k^2 u, \tag{5.1.14}$$

where the series $\sum_{k=1}^{\infty} a_k \partial_k^2 u$ converges strongly in L_p.

Proof. First we note that the solution of equation (5.1.12) has the form $u = -R^\lambda f$, where (R^λ) is the resolvent of semigroup (P_t) (see Sec. 5.1.1). Therefore the case reduces to a proof of inclusion $R^\lambda f \in W_p^{k+2}$. It is obviously sufficient to consider the case $k = 0$, $W_p^0 := L_p$.

1. For a function $\varphi \in \mathscr{D}$ we have

$$\partial_k R^\lambda f(\varphi) = -\langle R^\lambda f, \partial_k \varphi \rangle = -\langle f, \partial_k R^\lambda \varphi \rangle.$$

We shall use the Hölder inequality and Theorem 5.1.5.

$$|\partial_k R^\lambda f(\varphi)| \leq c_q \|f\|_{L_p} \|B R^\lambda \varphi\|_{L_q}$$

From the boundedness of operator BR^λ in L_q it now follows that linear functional $\partial_k R^\lambda f(\cdot) : \mathscr{D} \to \mathbb{R}^1$ admits a continuous extension to L_q. Therefore there exists a function $u_1 \in L_p$ such that

$$\partial_k R^\lambda f(\varphi) = \langle u_1, \varphi \rangle.$$

Thus, $R^\lambda f \in W_p^1$. The proof of inclusion $R^\lambda f \in W_p^2$ is based on the relation $B^2 = -\mathscr{L}$ and is similar to the previous one.

2. For any function $s \in L_q$ and number $n > m$ we shall have

$$\langle s, \sum_{k=m}^{n} a_k \partial_k^2 u \rangle = \langle s, \sum_{k=m}^{n} R_k^2 \mathcal{L}_p u \rangle = \langle s, \sum_{k=m}^{n} R_k g_k \rangle =$$

$$= \sum_{k=m}^{n} \langle R_k^* s, g_k \rangle = -\sum_{k=m}^{n} \langle R_k s, R_k \mathcal{L}_p u \rangle = -\langle \vec{R}_{mn} s, \vec{R}_{mn} \mathcal{L}_p u \rangle$$

where $\vec{R}_{mn} := (0, \ldots, R_m, \ldots, R_n, \ldots)$. Now applying Hölder's inequality and Theorem 5.1.7, we obtain

$$\left| \langle s, \sum_{k=m}^{n} a_k \partial_k^2 u \rangle \right| \leq ||\vec{R}_{mn} s||_{\vec{L}_q} ||\vec{R}_{mn} \mathcal{L}_p u||_{\vec{L}_p} \leq$$

$$\leq c_q ||s||_{L_q} ||\vec{R}_{mn} \mathcal{L}_p u||_{\vec{L}_p}.$$

Therefore for $m, n \to \infty$, we get

$$\left\| \sum_{k=m}^{n} a_k \partial_k^2 u \right\|_{L_p} \leq c_q ||\vec{R}_{mn} \mathcal{L}_p u||_{\vec{L}_p} \to 0,$$

the proof is finished. $\qquad\qquad\square$

Remark 5.1.4. For a vector $x = (x_i)_1^\infty \in \mathbb{R}^\infty$ and function $u \in W_p^1$ we set

$$\partial_x u := \sum_{k=1}^{\infty} x_k \partial_k u.$$

The function $\partial_x u$ (if it is in L_p) would naturally be called the L_p-derivative of u along the direction of vector x.

We consider the Hilbert space

$$H_A := \{x = (x_i)_1^\infty : |x|_A^2 = \sum_{k=1}^{\infty} \frac{x_k^2}{a_k} < \infty\}.$$

Based on Theorems 5.1.5 and 5.1.7 it is easy to show that in addition to Theorem 5.1.8, the following assertion holds: every weak solution u of equation (5.1.12) with right-hand side $f \in L_p$ has L_p-derivatives of first and second order along H_A, and $\forall x, y \in H_A$:

1) $||\partial_x u||_{L_p} \leq c(p, \lambda)|x|_A (||\mathcal{L}_p u||_{L_p} + ||u||_{L_p})$,

2) $||\partial_{x,y}^2 u||_{L_p} \leq c(p, \lambda)|x|_A |y|_A (||\mathcal{L}_p u||_{L_p} + ||u||_{L_p})$.

5.2 Weak solutions of elliptic equations (L_p-theory)

In this section we widen the concept of a weak solution of equation (5.1.12). For this we shall need to introduce several new definitions.

5.2.1

Let V be an open subset of $\mathbb{T}^\infty$ and $\mathcal{D}(V)$ a set of functions $\varphi \in \mathcal{D}$ such that $\operatorname{supp}\varphi \subset V$. For $1 < p < \infty$ and $k = 0, 1, \ldots$ we set

$$L_{p\,\mathrm{loc}}(V) = \{f : \forall \alpha \in \mathcal{D}(V),\quad \alpha f \in L_p\},$$
$$W^k_{p\,\mathrm{loc}}(V) = \{f : \forall \alpha \in \mathcal{D}(V),\quad \alpha f \in W^k_p\}.$$

We shall say that a function $u \in L_{p\,\mathrm{loc}}(V)$ is a weak solution of equation (5.1.12) on V if $\forall \varphi \in \mathcal{D}(V)$, equation (5.1.13) is satisfied.

In justification of the term "weak solution" we note that every function $u \in \mathcal{H}_{\varphi^\lambda}(V)$ is a weak solution of homogeneous equation (5.1.12) on V, and Green's potential $R^\lambda_{T^\infty\backslash V}(f)$ (see Sec. 2.7) is a weak solution of nonhomogeneous equation (5.1.12) on V with right-hand side $-f$.

The main result of this section is contained in the following theorem.

Theorem 5.2.1. *Let a function u be a weak solution of equation (5.1.12) on V. If the right-hand side f of this equation belongs to $W^k_{p\,\mathrm{loc}}(V)$, then $u \in W^{k+2}_{p\,\mathrm{loc}}(V)$,*

$$\mathcal{L}u = \sum_{k=1}^\infty a_k \partial_k^2 u,$$

where the series converges strongly in $L_p(V_0)$ for any open set $V_0 : \overline{V}_0 \subset V$.

Remark 5.2.1. In correspondence with Remark 5.1.4, Theorem 5.2.1 can be supplemented by the following assertion: the function u has $L_p(V_0)$-derivatives of first and second order along Hilbert space $H_A \subset \mathbb{R}^\infty$, where the mappings

$$x \to \partial_x u, \quad (x, y) \to \partial^2_{x,y} u$$

are respectively the bounded linear and bounded bilinear operators from H_A to $L_p(V_0)$.

5.2.2

In the process of proving Theorem 5.2.1 we shall use the scale of Banach spaces $\{D^s_p : s = 0, \pm 1, \ldots\}$, which we define in the following manner:

$$D^s_p := \{f \in \mathcal{D}^* : (B - I)^s f \in L_p\}, \quad \|f\|_s := \|(B - I)^s f\|_{L_p}.$$

We note that $B : \mathcal{D} \to \mathcal{D}$, and therefore operator $B : \mathcal{D}^* \to \mathcal{D}^*$ is correctly defined by the relation

$$Bf(\varphi) := f(B\varphi) \ (f \in \mathcal{D}^*, \varphi \in \mathcal{D}).$$

Proposition 5.2.1. *The following properties hold:*
1) D_p^s *is a Banach space; $\mathcal{D}$ is dense in D_p^s;*
2) *if $s_1 > s_2$, then $D_p^{s_1} \subset D_p^{s_2}$;*
3) $(D_p^s)^* = D_q^{-s}$ *($1/p + 1/q = 1$); the indicated duality is realized by bilinear functional $\langle u, v \rangle_0 = \langle (B - I)^s u, (B - I)^{-s} v \rangle$;*
4) *if $f \in D_p^s$, then $\partial_k f \in D_p^{s-1}$ ($k = 1, 2, \ldots$);*
5) *if $f \in D_p^s$ and $\alpha \in \mathcal{D}$, then $\alpha f \in D_p^s$;*
6) *if $f \in D_p^s$ then $R^\lambda f \in D_p^{s+2}$;*
7) *if $s \geq 0$, then $D_p^s \subset W_p^s$, $D_p^2 = \mathrm{dom}(\mathcal{L}_p)$.*

The proof of these properties is carried out the same way as in the finite-dimensional case (see for example [82]), but with the use of the infinite-dimensional *apriori* estimate from Theorem 5.1.5. We shall show, for example, that property 4) holds. We shall use the boundedness of operator $B(B-I)^{-1}$ in L_p and Theorem 5.1.5 and write the chain of inequalities for $f \in \mathcal{D}$:

$$||\partial_k f||_{s-1} = ||(B - I)^{s-1}\partial_k f||_{L_p} = ||\partial_k (B - I)^{-1}(B - I)^s f||_{L_p} \leq$$
$$\leq c_p ||B(B - I)^{-1}(B - I)^s f||_{L_p} \leq c_p^1 ||(B - I)^s f||_{L_p} = c_p^1 ||f||_s.$$

$$\square$$

Proof of Theorem 5.2.1. Let $\alpha \in \mathcal{D}(V)$. Setting $\tilde{u} = \alpha u$, we shall have $\tilde{u} \in L_p$ and

$$\mathcal{L}\tilde{u} = (\mathcal{L}\alpha)u + \lambda \tilde{u} + \alpha f + 2\nabla_A \alpha \cdot \nabla_A u.$$

The first three terms belong to $L_p = D_p^0$, and the last, by property 4) of Proposition 5.2.1 and relation

$$\beta \partial_s u = \partial_s (\beta u) - (\partial_s \beta)u, \ \beta \in \mathcal{D},$$

belongs to D_p^{-1}. Setting

$$g = (\mathcal{L}\alpha)u + \alpha f + 2\nabla_A \alpha \cdot \nabla_A u$$

we get that

$$\mathcal{L}\tilde{u} - \lambda \tilde{u} = g, \ g \in D_p^{-1}.$$

Therefore by property 6) of Proposition 5.2.1, we conclude that $\tilde{u} = -R^\lambda g \in D_p^1$. But then by property 4)

$$g \in D_p^0 = L_p = W_p^0.$$

Now, using Theorem 5.1.8, we conclude that $\tilde{u} \in W_p^2$ and series $\mathcal{L}\tilde{u} = \sum_{k=1}^{\infty} a_k \partial_k^2 \tilde{u}$ converges in L_p. Thus, the theorem is proved for the case $k = 0$. The general case is obtained from this case by induction on k. $\qquad\square$

5.2.3

Theorem 5.2.1 applied to function u such that $u \in \mathcal{H}_{\mathcal{L}\lambda}(V)$ shows that u has derivatives (in the sense of L_p) of every order. A similar conclusion with respect to ordinary derivatives, generally speaking, cannot be made: the function u can even turn out to be discontinuous (see Remark 4.4.1). However, if $X \in \mathcal{B}$ (see Sec. 3.2), then every function $u \in \mathcal{H}_{\mathcal{L}\lambda}(V)$ has continuous derivatives of every order. This property follows from the following theorem.

Theorem 5.2.2. *The following assertions are equivalent:*

1) Every weak solution of a homogeneous equation coincides almost everywhere with respect to Haar measure with some continuously differentiable function;

2) the process X is in class $\mathcal{B}$ (see Sec. 3.2).

Proof. 2) $\Rightarrow$1). Let u be a weak solution of the homogeneous equation on V, V_0 an open subset of V such that $\overline{V}_0 \subset V$; $\alpha \in \mathcal{D}(V)$ such that $\alpha|_{V_0} = 1$. We set $\tilde{u} = \alpha u$. According to Theorem 5.2.1:

$$\mathcal{L}\tilde{u} - \lambda\tilde{u} = g, \ g \in L_p, \ \mathrm{supp}\,g \subset T^{\infty}\backslash V_0.$$

Therefore $\tilde{u} = -R^{\lambda}g$ $(dx$-a.e.$)$. We set

$$g^+ = \max(g, 0), \ g^- = \max(-g, 0),$$
$$u^+ = R^{\lambda}g^+, \ u^- = R^{\lambda}g^-.$$

Then $\tilde{u} = u^- - u^+$ $(dx$-a.e.$)$. For every $n \geq 1$, the function $u_n^+ = R^{\lambda}(\min(g^+, n))$ is continuous and λ is harmonic on V_0. From the fact that $u_n^+ \uparrow u^+$ and the Brelot convergence property (see Corollary 3.2.1) it follows that the function u^+ is continuous on V_0. The continuity of function u^- on V_0 is proved analogously. Thus, it is proved that the function u coincides $(dx$-a.e.$)$ on V_0 with continuous λ-harmonic function $u^* = u^- - u^+$. Now applying this argument for every $k = 1, 2, \ldots$ to the weak solution $\partial_k u$ we arrive at the truth of assertion 1).

1) $\Rightarrow$ 2). Let $E^* = \mathbb{R}^1 \times \mathbb{T}^{\infty}$, γ the natural homomorphism of group E^* onto group $E = \mathbb{R}^1/2\pi\mathbb{Z}^1 \times \mathbb{T}^{\infty} \approx \mathbb{T}^{\infty}$ and X^* a space-homogeneous process on group E^* such that $\gamma(X^*) = X$. It is obvious that properties 1) and 2) are satisfied or are not satisfied for processes X^* and X simultaneously. Therefore we shall carry out the arguments for process X^*. Further, without loss of generality, we shall assume that in the sequence of coefficients of operator $\mathcal{L}$ the first coefficient equals one and consequently this sequence has the form $\{1, a_1, a_2, \ldots\}$. For convenience we shall

also denote elements of group E^* as (t, x) where $t \in \mathbb{R}^1$, and $x = (x_i)_1^\infty \in \mathbb{T}^\infty$. Thus the operator $\mathscr{L}$ has the form

$$\mathscr{L} = \partial_t^2 + \sum_{k=1}^\infty a_k \partial_k^2.$$

From this it follows that $X^* = Y \times Z$, where Y is a Wiener process on the line $\mathbb{R}^1$, and Z is the space-homogeneous process on group $\mathbb{T}^\infty$ associated with operator $\sum_{k=1}^\infty a_k \partial_k^2$.

We shall consider the special domain $V =]0, +\infty[\times \mathbb{T}^\infty \subset E^*$, $\partial V = \{0\} \times \mathbb{T}^\infty \approx \mathbb{T}^\infty$. It is easy to show (for example, with the help of the Fourier transform technique; see also [79, 80]), that for any bounded Borel function φ on $\mathbb{T}^\infty$ ($\approx \partial V$) the function

$$u(t, x) = Q_t \varphi(x), \quad (t, x) \in V,$$

where (Q_t) is the Cauchy semigroup associated with process Z (see Sec. 5.1.1), is harmonic with respect to process X^* on V. Therefore without difficulty we get that for any function $\varphi \in L_p(\mathbb{T}^\infty)$ the function $u(t, x) = Q_t \varphi(x)$ is a weak solution of equation $\mathscr{L}u = 0$ on set V.[1]

Now suppose assertion 1) holds; then there exists a continuous function $u^*(t, x)$ on V such that $u = u^*$ on V except for a set $dt \otimes dx$ of measure zero. Therefore without difficulty we deduce that for every $(t, x) \in V$

$$Q_t \varphi(x) = u^*(t, x).$$

In particular, from this it follows that for every $t > 0$

$$Q_t : L_p(\mathbb{T}^\infty) \to L_\infty(\mathbb{T}^\infty)$$

is a bounded linear operator. Every such operator, according to Buchwalow's theorem [70: Chapter 11, §1, Sec. 1.6, Corollary], is an integral operator with kernel $q_t(x, y)$ such that

$$\| \, \| \, q_t(x, y) \, \|_{y, L_q} \|_{x, L_\infty} < \infty, \quad \frac{1}{p} + \frac{1}{q} = 1.$$

Considering the permutability of operator Q_t with the shift operator, we conclude that

$$Q_t \varphi(x) = \int_{\mathbb{T}^\infty} q_t(x - y) \varphi(y) \, dy, \quad q_t(x) \in L_q.$$

If $1 < p \leq 2$, then $q \geq 2$ and consequently $q_t \in L_2$ ($\forall t > 0$). But then from the relation

$$\widehat{q}_t(\theta) = \exp(-t \sqrt{\Psi(\theta)}), \quad \theta \in \mathbb{Z}^{(\infty)},$$

---

1   For simplicity we shall consider the equation $\mathscr{L}u = 0$; in the case of equation $\mathscr{L}u - \lambda u = 0$ one must consider the Cauchy semigroup for process Z^λ.

it follows that $\widehat{q}_{2t} \in L_1(\mathbb{Z}^{(\infty)})$ and therefore function q_{2t} is continuous for all $t > 0$. If $p \geq 2$, then $1 < q \leq 2$. By the Hausdorff–Young theorem, $\widehat{q}_t \in L_p(\mathbb{Z}^{(\infty)})$, but then from the relation $|\widehat{q}_t|^p = \widehat{q}_{pt}$ it follows that $\widehat{q}_{pt} \in L_1(\mathbb{Z}^{(\infty)})$. Consequently the function q_{pt} is continuous for every $t > 0$.

Thus, for any $1 < p < \infty$, condition 1) implies that the semigroup of measures (q_t) associated with the Gaussian semigroup of process Z (see Sec. 5.1.1) consists of absolutely continuous measures with respect to Haar measure, having, furthermore, continuous densities. According to Lemma 4.3.2, this implies that $Z \in \mathfrak{B}$, and therefore in turn, it also follows that $X^* = Y \times Z \in \mathfrak{B}$. $\square$

5.3 Weyl's lemma and the hypoelliptic property

In this section we widen the concept of a weak solution in the following manner. Let μ and ν be measures on $\mathbb{T}^\infty$ and V an open subset of $\mathbb{T}^\infty$. We shall say that the measure μ is a weak solution of the equation

$$\mathscr{L}\mu - \lambda\mu = \nu \tag{5.3.1}$$

on the set V if for any function $\varphi \in \mathscr{D}(V)$, where $\mathscr{D}(V)$ is the set of functions $\varphi \in \mathscr{D}$ such that $\mathrm{supp}\varphi \subset V$, the following relation is satisfied:

$$\langle \mu, \mathscr{L}\varphi - \lambda\varphi \rangle = \langle \nu, \varphi \rangle. \tag{5.3.2}$$

5.3.1

First we shall consider some examples illustrating the concept of a weak solution. We shall consider the sequence $\{a_k\}_1^\infty$ such that $a_k = k^\alpha$, $\alpha \geq 0$, $k = 1, 2, \ldots$. Let (μ_t) be the corresponding Gaussian semigroup of measures and (r^λ) its resolvent. We shall fix $\lambda > 0$ and consider measure r^λ. As already noted (see Remarks 4.2.1 and 4.2.2), measure r^λ is a weak solution of equation $\mathscr{L}u - \lambda u = -\varepsilon_0$ on $V = \mathbb{T}^\infty$. In particular, measure r^λ is a weak solution of the homogeneous equation on $V = \mathbb{T}^\infty \backslash \{0\}$.

1. Let $\alpha = 0$. In this case, $a_k = 1$, $k = 1, 2, \ldots$ and consequently, $\mathscr{L} = \sum_{k=1}^\infty \partial_k^2$ is the infinite-dimensional Laplace operator. We shall show, following [33: Proposition 7] (see also [95] and [18]) that r^λ is singular with respect to Haar measure.

According to (4.3.2), $\mu_t = \otimes_1^\infty n_t$, where n_t is the Gaussian measure on $\mathbb{T}^1$ with parameters $(0, 2t)$. We have: $n_t(dx) = n_t(x)\,dx$,

$$n_t(x) = 2\pi(4\pi t)^{-1/2} \sum_{k \in \mathbb{Z}} \exp\left(-\frac{(x - 2\pi k)^2}{4t}\right).$$

For $k = 1, 2, \ldots$ and $t > 0$, we define the set:

$$A_{kt} := \left\{ x \in T^\infty : \prod_{i=1}^{k} n_t(x_i) > 1 \right\}.$$

We denote the Haar measure of set A_{kt} by $|A_{kt}|$. We have

$$|A_{kt}| = \int_{\pi_{k\infty}(A_{kt})} dx_1 \ldots dx_k \leq \int_{\mathbb{T}^k} \left(\prod_{i=1}^{k} n_t(x_i) \right)^{1/2} dx_1 \ldots dx_k = \rho(t)^k,$$

where $\rho(t) = \int_{\mathbb{T}^1} \sqrt{n_t(x)}\, dx$, $0 < \rho(t) < 1$. Let $A_t = \cap_{n=1}^{\infty} \cup_{k=n}^{\infty} A_{kt}$; then A_t is a Borel set and

$$|A_t| = \lim_{n \to \infty} | \cup_{k=n}^{\infty} A_{kt} | \leq \lim_{n \to \infty} \sum_{k=n}^{\infty} \rho(t)^k = 0.$$

We fix $t > 0$; since $n_s(x)/n_t(x) \to 1$ for $s \to t$ uniformly with respect to $x \in \mathbb{T}^1$, then there exists a neighborhood $V_t \subset\,]0, +\infty[$ of point t such that for $s \in V_t$, $x \in \mathbb{T}^1$,

$$n_s(x)/n_t(x) \leq \rho(t)^{-1/2}.$$

For $s \in V_t$ we have

$$\mu_s(\mathbb{T}^\infty \backslash A_{kt}) = \int_{\mathbb{T}^k \backslash \pi_{k\infty}(A_{kt})} \prod_{i=1}^{k} n_s(x_i)\, dx_1 \ldots dx_k \leq$$

$$\leq \int_{\mathbb{T}^k \backslash \pi_{k\infty}(A_{kt})} \rho(t)^{-k/2} \prod_{i=1}^{k} n_t(x_i)\, dx_1 \ldots dx_k \leq$$

$$\leq \rho(t)^{-k/2} \int_{\mathbb{T}^k \backslash \pi_{k\infty}(A_{kt})} \prod_{i=1}^{k} [n_t(x_i)]^{1/2} dx_1 \ldots dx_k \leq \rho(t)^{k/2},$$

from which it follows that $\mu_s(\mathbb{T}^\infty \backslash A_t) = 0$. The system of sets (V_t) forms a cover of interval $]0, +\infty[$; therefore there exists a sequence (t_k) such that

$$\cup_{k=1}^{\infty} V_{t_k} = \,]0, +\infty[\,.$$

The set $A = \cup_{k=1}^{\infty} A_{t_k}$ is Borel and $|A| = 0$. For any $s > 0$ there exists t_k such that $s \in V_{t_k}$, therefore

$$\mu_s(\mathbb{T}^\infty \backslash A) \leq \mu_s(\mathbb{T}^\infty \backslash A_{t_k}) = 0.$$

Therefore we conclude that

$$r^\lambda(\mathbb{T}^\infty \backslash A) = \int_0^\infty e^{-\lambda s} \mu_s(\mathbb{T}^\infty \backslash A)\, ds = 0,$$

i.e. r^λ is a singular measure.

2. Let $0 < \alpha \le 1$. By Theorem 4.3.2, measure r^λ is absolutely continuous with respect to Haar measure, and its excessive density $r^\lambda(x)$ is lower semicontinuous and strictly positive. We shall show that the set

$$A = \{x \in \mathbb{T}^\infty \backslash \{0\} : r^\lambda(x) = +\infty\}$$

is not empty and $x = 0$ is its limit point (see [32: Theorem 3.2]). For $x = (x_1, \ldots, x_k, 0, \ldots)$ and $0 < t < 1$ we shall have

$$\mu_t(x) = \prod_{s=1}^{k} n_{s^\alpha t}(x_s) \prod_{s=k+1}^{\infty} (1 + 2\varphi(s^\alpha t)) \ge$$

$$\ge \pi^{k/2} t^{-k/2} (k!)^{-\alpha/2} \exp\left(-\frac{1}{t} \sum_{s=1}^{k} \frac{x_s^2}{4s^\alpha}\right) \prod_{\substack{s=k+1, \\ s^\alpha t \le 1}}^{\infty} (1 + 2\varphi(s^\alpha t)) \ge$$

$$\ge A(k, \alpha) t^{-k/2} \exp\left(-\frac{1}{t} \sum_{s=1}^{k} \frac{x_s^2}{4s^\alpha}\right) (1 + 2\varphi(1))^{t^{-1/\alpha}} =$$

$$= A(k, \alpha) t^{-k/2} \exp\left(\frac{1}{t} \left[b \cdot t^{1-1/\alpha} - \sum_{s=1}^{k} \frac{x_s^2}{4s^\alpha}\right]\right),$$

$$b := \ln(1 + 2\varphi(1)) > 0.$$

If $\alpha = 1$, then for $x \in \{y \in \mathbb{T}^\infty \backslash \{0\} : \sum_{s=1}^{k} \frac{y_s^2}{4s^\alpha} < b\}$ we obtain

$$r^\lambda(x) = \int_0^\infty e^{-\lambda t} \mu_t(x) \, dt > \int_0^1 e^{-\lambda t} \mu_t(x) \, dt = +\infty.$$

If $0 < \alpha < 1$, then $r^\lambda(x) = +\infty$ for any $x = (x_1, \ldots, x_k, 0, 0, \ldots)$.

3. Let $\alpha > 1$. Then by Theorem 4.3.3, the function $r^\lambda(x)$ is continuous on $\mathbb{T}^\infty \backslash \{0\}$ and $r^\lambda(0) = +\infty$, and by Theorem 5.2.2 this function is continuously differentiable on $\mathbb{T}^\infty \backslash \{0\}$. Moreover, from Remark 5.2.1 and Theorem 5.2.2 it follows that $r^\lambda(x)$ on $\mathbb{T}^\infty \backslash \{0\}$ has continuous derivatives of every order along Hilbert space H_A:

$$H_A = \left\{x \in \mathbb{R}^\infty : \sum_{k=1}^{\infty} \frac{x_k^2}{a_k} < \infty\right\},$$

where for any $k \ge 1$ the mapping $(x, \ldots, y) \to \partial^k_{x, \ldots, y} r^\lambda$ is a bounded k-linear operator from H_A into $\mathbb{C}(V)$, for any open set V such that $\overline{V} \subset \mathbb{T}^\infty \backslash \{0\}$.

5.3.2

Following the classical definition, we shall say that the *hypoelliptic* property holds if every weak solution μ of equation $\mathcal{L}\mu - \lambda\mu = \nu$ on V is absolutely continuous with respect to Haar measure on V and has a density that is continuously differentiable any number of times whenever measure ν is absolutely continuous with respect to Haar measure on V and has a density that is continuously differentiable any number of times there.

Theorem 5.3.1. *The following assertions are equivalent:*
1) *the hypoelliptic property holds;*
2) *process X is in class $\mathcal{B}$ (see Sec. 3.2).*
3) $N(\lambda) = o(\lambda), \quad \lambda \to \infty$ *(see Sec. 4.3.4).*

The proof of this theorem is preceded by several auxiliary results.

5.3.3

Let $V \subset \mathbb{R}^n$ be an open set with compact closure and smooth boundary ∂V, $d\sigma$ be an element of area in ∂V, and $\mathcal{L}_n = \sum_{k=1}^n a_k \partial_k^2$. We fix a continuous function φ on ∂V and let u be a solution of the Dirichlet problem

$$\begin{cases} \mathcal{L}_n u - \lambda u = 0, \\ \quad u|_{\partial V} = \varphi. \end{cases}$$

It is well-known [82: Chapter 3, §14}] that within V the function u admits the representation

$$u(x) = \int_{\partial V} \varphi(y) P_V^\lambda(x, y) \, d\sigma(y).$$

Function $P_V^\lambda(x, y)$, $(x, y) \in V \times \partial V$, is called the λ-Poisson kernel of V. According to [30: Theorem 5.12] for any compact set $K \subset V$ there exist numbers $A, B > 0$ such that

$$\sup_{x \in K, y \in \partial V} P_V^\lambda(x, y) \le A e^{-B\sqrt{\lambda}}. \tag{5.3.3}$$

Let $V = B_r$ be the ball of radius r with center at zero. The following identity will be useful for us:

$$P_{B_1}^{\lambda r^2}(x, y) = P_{B_r}^\lambda(rx, ry) r^{n-1} \quad (x \in B_1, y \in \partial B_1) \tag{5.3.4}$$

5.3.4

We shall denote elements of group $\mathbb{T}^{n+m}$ by xz, where $x \in \mathbb{T}^n$ and $z \in \mathbb{T}^m$. We shall consider the cylindrical set $V_r := B_r \times \mathbb{T}^m$, where $r \le 1$, $B_r \subset \mathbb{T}^n$ ($\mathbb{T}^n \approx [-\pi, \pi]^n$). On $\partial V_r = \partial B_r \times \mathbb{T}^m$ we define the measure $d\sigma_r(xz) := d\sigma_r(x) \otimes dz$, where $d\sigma_r(x)$ is an element of area on ∂B_r, dz is the Haar measure on $\mathbb{T}^m$. Let u be the solution of the Dirichlet problem

$$\begin{cases} \mathcal{L}_{n+m} u - \lambda u = 0, \\ \quad u|_{\partial V_r} = \varphi. \end{cases} \tag{5.3.5}$$

According to [30: Theorem 8.3], within V_r the function u admits the representation

$$u(xz) = \int_{\partial V_r} \varphi(yt) P_{V_r}^{\lambda}(xz, yt) \, d\sigma_r(yt).$$

Function $P_{V_r}^{\lambda}(xz, yt)$ is called the λ-Poisson kernel of set V_r and has the form

$$P_{V_r}^{\lambda}(xz, yt) = \sum_{\theta \in \{0\} \times \mathbb{Z}^m \subset \mathbb{Z}^{n+m}} P_{B_r}^{\lambda + \Psi(\theta)}(x, y) \exp\{i\theta \cdot (xz - yt)\}, \tag{5.3.6}$$

where $\Psi(\theta) = \sum_{k=1}^{n+m} a_k \theta_k^2$.

Let $q_t^{n+m} = \int_0^{\infty} \mu_s^{n+m} h_t(s) \, ds$ be the Cauchy semigroup associated with Gaussian semigroup $\mu_t^{n+m} = \otimes_{k=1}^{n+m} n_{a_k t}$ (see Sec. 4.2.4). We set

$$D_{n,m,r} := A \left(\int_0^r \frac{s^{n-1} ds}{q_{sB}^{n+m}(0)} \right)^{-1},$$

where A and B are the constants from inequality (5.3.3) for $V = B_1$.

Lemma 5.3.1. *Let u be a solution of Dirichlet problem (5.3.5). The following inequality holds:*

$$|u(0)| \leq D_{n,m,r} \|u\|_{L_1(V_r)}.$$

Proof. For $s \leq r$ we write the equality

$$u(xz) = \int_{\partial V_s} P_{V_s}^{\lambda}(xz, yt) u(yt) \, d\sigma_s(yt).$$

Setting $xz = 0$ in this equality, we get

$$
\begin{aligned}
|u(0)| &\leq \max_{yt} P_{V_s}^{\lambda}(0, yt) \int_{\partial V_s} |u(yt)| \, d\sigma_s(yt) = \\
&= \max_{yt} P_{V_s}^{\lambda}(0, yt) \int_{\partial B_s} d\sigma_s(y) \int_{T^m} |u(yt)| \, dt.
\end{aligned}
$$

Now we use relations (5.3.3), (5.3.4), and (5.3.6),

$$
\begin{aligned}
\max_{yt} P_{V_s}^{\lambda}(0, yt) &\leq \sum_{\theta \in \mathbb{Z}^{n+m}} \max_y P_{B_s}^{\lambda + \Psi(\theta)}(0, y) = \\
&= \sum_{\theta \in \mathbb{Z}^{n+m}} s^{1-n} \max_y P_{B_1}^{(\lambda + \Psi(\theta))s^2}(0, y) \leq s^{1-n} \sum_{\theta \in \mathbb{Z}^{n+m}} A e^{-s\sqrt{\lambda + \Psi(\theta)}B} \leq \\
&\leq A s^{1-n} q_{sB}^{n+m}(0).
\end{aligned}
$$

Thus, the following inequality is established:

$$|u(0)| \frac{s^{n-1}}{q_{sB}^{n+m}(0)} \leq A \int_{\partial B_s} d\sigma_s(y) \int_{T^m} |u(yt)| \, dt.$$

Integrating this inequality with respect to $s \in \,]0, r[$, we obtain the desired result. $\qquad\square$

5.3.5

Let V be an open subset of $\mathbb{T}^\infty$ and $\mathcal{F}$ the set of λ-harmonic functions on V such that

$$\sup\{||u||_{L_1(V)} : u \in \mathcal{F}\} < \infty.$$

Lemma 5.3.2. *If process X is in class $\mathcal{B}$, then for any compact set $K \subset V$ the set $\mathcal{F}_K = \{u|_K : u \in \mathcal{F}\}$ is a relatively compact subset of $\mathbb{C}(K)$.*

Proof. Without loss of generality, we may assume that $V = V_n \times \mathbb{T}^\infty$ and $K = K_n \times \mathbb{T}^\infty$ for some $K_n \subset V_n \subset \mathbb{T}^n$. We choose $r > 0$ small enough such that $\cup_{x \in K_n}\{x + B_r\} \subset V_n$. Let ν_{n+m} be the Haar measure of closed subgroup $\{0\} \times \mathbb{T}^\infty$, where $\{0\} \subset \mathbb{T}^{n+m}$. For function $u \in \mathcal{F}$ we form the sequence $\{u_n\}$ of cylindrical λ-harmonic functions: $u_m = u * \nu_{n+m}$. We apply Lemma 5.3.1 to the function

$$u_{m,x}(z) := u_m(x + z) \ (x \in K)$$

assuming that $z \in B_r$:

$$|u_m(x)| = |u_{m,x}(0)| \le D_{n,m,r}||u_{m,x}||_{L_1(B_r \times \mathbb{T}^m)} =$$
$$= D_{n,m,r}||u_m||_{L_1(\{x+B_r\} \times \mathbb{T}^m)} \le$$
$$\le D_{n,m,r}||u_m||_{L_1(V_n \times \mathbb{T}^m)} \le D_{n,m,r}||u||_{L_1(V)}.$$

Let (q_t) be the Cauchy semigroup associated with Gaussian semigroup (μ_t) of process X (see Sec. 5.1.1). We have

$$D_{n,m,r} = A\left(\int_0^r \frac{s^{n-1}ds}{q_{sB}^{n+m}(0)}\right)^{-1} \uparrow A\left(\int_0^r \frac{s^{n-1}ds}{q_{sB}(0)}\right)^{-1} := D \ (m \uparrow \infty).$$

The sequence $u_m = u * \nu_{n+m}$ converges uniformly in K to the function u, and therefore

$$\sup_K |u| = \lim_{m \to \infty} \sup_K |u_m| \le D||u||_{L_1(V)}.$$

Now applying this inequality to functions $u \in \mathcal{F}$, we conclude that

$$\sup\{||u||_{\mathbb{C}(K)} : u \in \mathcal{F}_K\} < \infty.$$

Thus the set $\mathcal{F}$ of λ-harmonic functions is uniformly bounded on K, but then this set is also equicontinuous on K (see [46: Theorem 11.1.1], [97]). Consequently $\mathcal{F}_K$ is a relatively compact subset of $\mathbb{C}(K)$. $\qquad\square$

5.3.6

Proof of Theorem 5.3.1. We note that assertion 1) $\Rightarrow$ 2) is a particular case of assertion 1) $\Rightarrow$ 2) of Theorem 5.2.2; therefore we consider 2) $\Rightarrow$ 1). Let W

be an open subset of V and g a continuously differentiable function such that $\mathrm{supp}\, g \subset V$ and $d\nu(x) = g(x)dx$ on W. From the permutability of the operators of differentiation and convolution it follows that the function $R^\lambda g$ is continuously differentiable.

Measure $\mu^1 := -R^\lambda g(x)dx$ is a weak solution of equation $\mathscr{L}\mu^1 - \lambda\mu^1 = \nu$ on W; therefore measure $\mu^0 := \mu - \mu^1$ is a weak solution of equation $\mathscr{L}\mu^0 - \lambda\mu^0 = 0$ on W. We shall show that measure μ^0 is absolutely continuous with respect to Haar measure on W and has continuously differentiable density; then the proof of assertion 2)$\Rightarrow$1) of the theorem will be finished.

Without loss of generality, we may assume that $W = W_n \times \mathbb{T}^\infty$, $W_n \subset \mathbb{T}^n$. Let $\mathscr{B}_{n+m}$ be the σ-algebra of cylindrical Borel subsets of $\mathbb{T}^\infty$ with bases in $\mathbb{T}^{n+m}$. We shall denote by μ^0_{n+m} the restriction of μ^0 to σ-algebra $\mathscr{B}_{n+m}$. Measure μ^0_{n+m} is a weak solution of the equation $\mathscr{L}_{n+m}\mu^0_{n+m} - \lambda\mu^0_{n+m} = 0$ on set $W_n \times \mathbb{T}^m \subset \mathbb{T}^{n+m}$. By the classical Weyl lemma (see, for example, [44], [77: 4.2]) the measure μ^0_{n+m} is absolutely continuous with respect to Haar measure on $W_n \times \mathbb{T}^m$ and has continuously differentiable density $\mu^0_{n+m}(x)$. Cylindrical function $\mu^0_{n+m}(x)$ is λ-harmonic on W, where for any compact set $K \subset W$

$$\|\mu^0_{n+m}\|_{L_1(K)} \leq \mathrm{var}\, \mu^0(K) < \infty.$$

By Lemma 5.3.2, the set $\mathscr{F} = \{\mu^0_{n+m}\}$ is relatively compact in $\mathbb{C}(K)$; let μ^0_∞ be one of its limit points. On the other hand, sequence $\mathscr{F}$ weakly converges to measure μ^0; therefore $d\mu^0(x) = \mu^0_\infty(x)dx$. Continuous function $\mu^0_\infty(x)$ is a weak solution of the homogeneous equation, and by Theorem 3.2.2 it is continuously differentiable.

2) $\Longleftrightarrow$ 3): see the proof of Theorem 4.3.3. $\qquad\qquad\qquad\qquad\square$

5.3.7

We shall assume that process X is in class $\mathscr{B}$ (see Sec. 3.2), and consequently the hypoelliptic property holds. Up to now we have considered differentiation of functions with respect to any finite collection of variables, i.e. differentiation along finite-dimensional subgroups. Here we shall show that some infinite-dimensional subgroups are also admissible.

We set

$$H_A = \left\{ x = (x_i)_1^\infty \in \mathbb{R}^\infty : |x|_A^2 := \sum_{k=1}^\infty \frac{x_k^2}{a_k} < \infty \right\}$$

and let γ be the natural homomorphism of group $\mathbb{R}^\infty$ onto group $\mathbb{T}^\infty = \mathbb{R}^\infty/2\pi\mathbb{Z}^\infty$. For $x \in H_A$ and $t \in \mathbb{R}^1$ we set $x_t = \gamma(t \cdot x)$. It is obvious that $(x_t)_{t \in \mathbb{R}^1}$ is a one-parameter subgroup of $\mathbb{T}^\infty$.

We shall say that a function u defined on some neighborhood of the point $x_0 \in \mathbb{T}^\infty$ is differentiable at x_0 in the direction of vector $x \in H_A$ if the function

$t \to u(x_0 + x_t)$ is differentiable at point $t = 0$. We shall denote the corresponding derivative by $\partial_x u(x_0)$ and call it the derivative along the direction of vector x. The higher derivatives $\partial^2_{x,y} u$, $\partial^3_{x,y,z} u$, ... are defined analogously.

We denote by $\mathbb{C}^k_{\mathscr{L}}(V)$ $(1 \leq k \leq \infty)$ the set of functions defined on an open set V which are continuously differentiable up to k-th order inclusive along the direction of any vector $x \in H_A$.

Theorem 5.3.2. *Let u be a weak solution of equation $\mathscr{L}u - \lambda u = g$ on V. If $g \in \mathbb{C}^\infty_{\mathscr{L}}(V)$, then also $u \in \mathbb{C}^\infty_{\mathscr{L}}(V)$.*

Proof. As in Theorem 5.3.1, the proof reduces to the case when u is a weak solution of the homogeneous equation, i.e. is a λ-harmonic function. Further arguments can be carried out based on Remark 5.2.1 and Theorem 5.2.2, as was already done in the analysis of Example 3 in Sec. 5.3.1. However, here we carry out an argument based on the representation of function u in the form of a Poisson integral (see Sec. 5.3.4).

First let function u be cylindrical; then the existence of a derivative $\partial_x u$ and the equality

$$\partial_x u = \sum x_k \partial_k u := (x, \nabla)u$$

are well-known facts from finite-dimensional analysis. In addition, from the representation of function u in the form of a Poisson integral, the inequality

$$\sup_{y \in V_r} |\partial_x u(x_0 + y)| \leq c_r |x|_A \sup_{y \in \partial V_{2r}} |u(x_0 + y)| \tag{5.3.7}$$

can easily be deduced, in which $x_0 \in V$, V_r is a cylindrical set whose base is the ball of radius $r < 1$ with center at zero, and c_r is a constant not depending on the dimension.

Every λ-harmonic function u on V can be uniformly approximated on V_{2r} by cylindrical harmonic functions. Inequality (5.3.7) now shows that bilinear functional $\partial_x u$ can be extended with respect to continuity up to a bounded bilinear functional on $H_A \times \mathscr{H}_{\mathscr{L}\lambda}(V_{2r})$. Therefore in the standard manner we can also deduce the differentiability of any function $u \in \mathscr{H}_{\mathscr{L}\lambda}(V)$ along the direction $x \in H_A$ and the equality $\partial_x u = (x, \nabla)u$. Finally we note that the function $y \to \partial_x u(y)$ again is a weak solution of the homogeneous equation on V, and therefore further differentiation is possible. $\square$

Remark 5.3.1. According to Theorem 5.3.1, every function $u \in \mathscr{H}_{\mathscr{L}\lambda}(V)$ is continuously differentiable, $\mathscr{L}u - \lambda u = 0$, where by Theorem 5.2.1, the series $\mathscr{L}u = \sum_{k=1}^\infty a_k \partial^2_k u$ converges in $L_p(V_0)$, for any open set $V_0 : \overline{V}_0 \subset V$.

We shall show that if $X \in \mathscr{B}$, then the series $\sum_{k=1}^\infty a_k \partial^2_k u$ converges uniformly on every compact set $K \subset V$. In fact, the sequence of λ-harmonic functions $u_n = \sum_{k=1}^n a_k \partial^2_k u$ converges in $L_p(V_0)$ to function λu. Consequently this sequence is bounded in $L_p(V_0)$, and thus is bounded also in $L_1(V_0)$. By Lemma 5.3.2 this sequence is relatively compact in $\mathbb{C}(K)$, for any compact set $K \subset V_0$. In accordance with this, λu is the unique limit point of this sequence.

5.4 Bessel potentials on group $\mathbb{T}^\infty$

For any $\beta > 0$, we set

$$\mathcal{J}_\beta(dx) = \frac{1}{\Gamma\left(\frac{\beta}{2}\right)} \int_0^\infty e^{-t} t^{\frac{\beta}{2}-1} \mu_t(dx)\, dt.$$

We have: $\mathcal{J}_2(dx) = r^1(dx)$, where (r^λ) is the resolvent of Gaussian semigroup (μ_t), and $\widehat{\mathcal{J}}_\beta(\theta) = (1 + \Psi(\theta))^{-\frac{\beta}{2}}$, where $\Psi(\theta) = \sum_{k=1}^\infty a_k \theta_k^2$. The family of measures $(\mathcal{J}_\beta)_{\beta>0}$ has the semigroup property with respect to convolution $\mathcal{J}_\alpha * \mathcal{J}_\beta = \mathcal{J}_{\alpha+\beta}$ and for $\beta \to 0$ the weak convergence of measures $\mathcal{J}_\beta \to \varepsilon_0$ takes place.

By analogy with the classical case [101: Chapter 5] the Bessel potential $\mathcal{J}_\beta[f]$ of function f is defined by the equality: $\mathcal{J}_\beta[f] = \mathcal{J}_\beta * f$. We shall investigate the properties of differentiability of Bessel potentials $\mathcal{J}_\beta[f]$ for functions $f \in \mathbb{C}(\mathbb{T}^\infty)$ and, most importantly, properties of differentiability of potential $\mathcal{J}_2[f] = R^1 f$. This potential, as already noted in Sec. 5.3, is a weak solution of the Poisson equation $\mathcal{L}u - u = -f$. We shall show that under fulfillment of certain conditions on coefficients $\{a_k\}_1^\infty$ of operator $\mathcal{L} = \sum_{k=1}^\infty a_k \partial_k^2$ and function f, the potential $\mathcal{J}_2[f]$ is a strong solution, i.e. twice continuously differentiable, where series $\mathcal{L}\mathcal{J}_2[f] = \sum_{k=1}^\infty a_k \partial_k^2 \mathcal{J}_2[f]$ converges absolutely and uniformly.

5.4.1

We shall begin with the properties of kernel $\mathcal{J}_\beta(dx)$, assuming that condition (N_1) (see Corollary 4.3.1) is satisfied.

The proof of the following proposition is similar to the proof of the first part of Lemma 4.3.1, and therefore we shall omit it.

Proposition 5.4.1. *Let condition (N_1) be satisfied. Then for any $\beta > 0$, the measure $\mathcal{J}_\beta(dx)$ is absolutely continuous with respect to Haar measure, and its density $\mathcal{J}_\beta(x)$ is continuous in the extended sense on $\mathbb{T}^\infty$, finite on set $\mathbb{T}^\infty \setminus \{0\}$ and $\mathcal{J}_\beta(0) = +\infty$.*

Let $n_t(dx)$ be the Gaussian measure on $\mathbb{T}^1$ ($\approx [-\pi, \pi]$) with parameters $(0, 2t)$ and $n_t(x)$ the density of this measure with respect to Haar measure ($=$ normed

Lebesgue measure on interval $[-\pi, \pi]$)

$$n_t(x) = \sqrt{\frac{\pi}{t}} \sum_{k\in\mathbb{Z}} \exp\left\{-\frac{(x-2\pi k)^2}{4t}\right\}.$$

Lemma 5.4.1. *The following inequality holds:*

$$|\partial_x(\ln n_t(x))| \le \frac{\pi}{t}\sin\frac{x}{2} \quad (t>0,\ 0\le x\le\pi).$$

Proof. We have $n_t(x) = \Theta_3\left(\frac{x}{2}, e^{-t}\right)$, where

$$\Theta_3(x, q) = \sum_{k\in\mathbb{Z}} q^{k^2} e^{2ikx},$$

is a theta-function. From Jacobi's formula for Θ_3 we get

$$n_t(x) = \prod_{k=1}^{\infty}(1 - e^{-2kt}) \prod_{k=1}^{\infty}(1 + 2e^{-(2k-1)t}\cos x + e^{-(4k-2)t}).$$

After the obvious transformations we get

$$\partial_x(\ln n_t(x)) = -\frac{1}{2}\sin x \sum_{k=0}^{\infty}\frac{1}{\sinh^2(k+\frac{1}{2})t + \cos^2\frac{x}{2}}.$$

Starting with this identity we carry out the further calculations

$$|\partial_x(\ln n_t(x))| \le \sin x \int_0^\infty \frac{dz}{\sinh^2 tz + \cos^2\frac{x}{2}} \le$$

$$\le \sin x \int_0^\infty \frac{dz}{t^2 z^2 + \cos^2\frac{x}{2}} = \frac{\pi\sin x}{2t\cos\frac{x}{2}} = \frac{\pi}{t}\sin\frac{x}{2},$$

which finishes the proof. $\qquad\qquad\square$

As in Sec. 5.1.3 (Theorem 5.1.5), let $\nabla_A = \left(\sqrt{a_k}\partial_k\right)_1^\infty$ and $\vec{L}_1 = \vec{L}_1(T^\infty, dx)$ be a Banach space of vector functions whose modulus is integrable with respect to Haar measure.

Lemma 5.4.2. *Let the following condition be satisfied:*

$$\lim_{k\to\infty}\frac{a_k}{k^\alpha} = \infty \quad (\alpha > 1). \qquad\qquad (N_\alpha)$$

Then for any $0 < \gamma < 1 - \frac{1}{\alpha}$ and for all $t > 0$ we have the inequality

$$\|\nabla_A \mu_t\|_{\vec{L}_1} \le \text{const}\cdot t^{\frac{\gamma}{2}-1}.$$

Proof. From the fact that $\mu_t = \otimes_{k=1}^\infty n_{a_k t}$ it follows that

$$\mu_t(x) = \prod_{k=1}^{\infty} n_{a_k t}(x_k), \quad x = (x_k)_1^\infty \in \mathbb{T}^\infty.$$

Applying the obvious transformations and Lemma 5.4.1, we get

$$\|\nabla_A \mu_t\|_{\vec{L}_1} \leq \frac{\pi}{t} \int_{\mathbb{T}^\infty} \sqrt{\sum_{k=1}^\infty \frac{1}{a_k} \sin^2 \frac{x_k}{2}} \, \mu_t(dx).$$

Let $n_t^0(dx) = (4\pi t)^{-1/2} \exp\left(-\frac{x^2}{4t}\right) dx$ be a Gaussian measure on $\mathbb{R}^1$ and $\mu_t^0(dx) = \otimes_{k=1}^\infty n_{a_k t}^0(dx_k)$ a Gaussian measure on $\mathbb{R}^\infty$. It is easy to see that the integral on the right-hand side of the obtained inequality can be written in the form of an integral with respect to measure $\mu_t^0(dx)$ (after periodic continuation of the integrand from set $\mathbb{T}^\infty = \mathbb{R}^\infty / 2\pi\mathbb{Z}^\infty$ to $\mathbb{R}^\infty$). Thus, we have

$$\|\nabla_A \mu_t\|_{\vec{L}_1} \leq \frac{\pi}{t} \int_{\mathbb{R}^\infty} \sqrt{\sum_{k=1}^\infty \frac{1}{a_k} \sin^2 \frac{x_k}{2}} \, \mu_t^0(dx) \leq$$

$$\leq \frac{\pi}{t} \sqrt{\int_{\mathbb{R}^\infty} \sum_{k=1}^\infty \frac{1}{a_k} \sin^2 \frac{x_k}{2} \mu_t^0(dx)} =$$

$$= \frac{\pi}{t} \sqrt{\sum_{k=1}^\infty \int_{\mathbb{R}^1} \frac{1}{a_k} \sin^2 \frac{x_k}{2} n_{a_k t}^0(dx_k)} = \frac{\pi}{t} \sqrt{\int_{\mathbb{R}^1} g(x) n_t^0(dx)}$$

From condition (N_α) it follows that the series $\sum_{k=1}^\infty \frac{1}{a_k^{1-\gamma}}$ converges, and therefore

$$g(x)/|x|^{2\gamma} = \sum_{k=1}^\infty \frac{1}{a_k} \sin^2 \frac{\sqrt{a_k}\,x}{2} / |x|^{2\gamma} =$$

$$= \sum_{k=1}^\infty \frac{\sin^{2\gamma} \frac{\sqrt{a_k}\,x}{2}}{(\sqrt{a_k}|x|)^{2\gamma}} \cdot \frac{\sin^{2-2\gamma} \frac{\sqrt{a_k}\,x}{2}}{a_k^{1-\gamma}} \leq \sum_{k=1}^\infty \frac{1}{2^{2\gamma}} \cdot \frac{1}{a_k^{1-\gamma}} < \infty.$$

From the obtained estimate for g we now obtain

$$\|\nabla_A \mu_t\|_{\vec{L}_1} \leq \text{const} \cdot t^{-1} \sqrt{\int_{\mathbb{R}^1} |x|^{2\gamma} n_t^0(dx)} = \text{const} \cdot t^{\frac{\gamma}{2}-1}.$$

The proof is finished. $\qquad\qquad\qquad\qquad\qquad\qquad\qquad\qquad\qquad\qquad\quad \square$

Proposition 5.4.2. *Let condition (N_α) be satisfied. Then for any $\beta > 1 + \frac{1}{\alpha}$ we have*

$$\|\nabla_A \mathcal{G}_\beta\|_{\vec{L}_1} < \infty.$$

Proof. We choose γ such that $\beta + \gamma > 2$. Then applying Lemma 5.4.2, we obtain

$$\|\nabla_A \mathcal{J}_\beta\|_{\bar{L}_1} \le \frac{1}{\Gamma\left(\frac{\beta}{2}\right)} \int_0^\infty e^{-t} t^{\frac{\beta}{2}-1} \|\nabla_A \mu_t\|_{\bar{L}_1}\, dt \le$$

$$\le \text{const} \int_0^\infty e^{-t} t^{\frac{\beta+\gamma}{2}-2}\, dt < \infty. \qquad \square$$

Remark 5.4.1. Let condition (N_1) be satisfied. We shall consider the sequence of positive numbers $B = (b_k)_1^\infty$ satisfying the condition: $\sum_{k=1}^\infty b_k/a_k < \infty$ and we set $\nabla_B = (\sqrt{b_k}\partial_k)_1^\infty$. Analysis of the proof of Lemma 5.4.2 shows that the following inequality holds:

$$\|\nabla_B \mu_t\|_{\bar{L}_1} \le \text{const} \cdot t^{-1/2}.$$

Repeating the arguments from Proposition 5.4.2, we get that the inequality $\|\nabla_B \mathcal{J}_\beta\|_{\bar{L}_1} < \infty$ is satisfied for any $\beta > 1$. Therefore for all $\beta > 1$ this implies, in particular, the continuous differentiability of kernel $\mathcal{J}_\beta(x)$ on the set $\mathbb{T}^\infty \backslash \{0\}$ and integrability of its derivatives $\partial_k \mathcal{J}_\beta(x)$, $k = 1, 2, \ldots$. These facts agree well with the known finite-dimensional results (see [101: Chapter 5]). We note in passing that the derivatives of higher order are not integrable functions.

Remark 5.4.2. In some cases the integrability of the function $|\nabla_B \mathcal{J}_\beta(x)|$ takes place, if sequence $(b_k)_1^\infty$ increases faster than sequence $(a_k)_1^\infty$. For example, let condition (N_α) be satisfied with indices $\alpha > 2$, and let $b_k = a_k \Delta_k$, where $\Delta_k = k \ln^2 k$ $(k \ge 2)$ and $\Delta_1 = 1$. As in the proof of Lemma 5.4.2, we get

$$\|\nabla_B \mu_t\|_{\bar{L}_1} \le \text{const} \cdot t^{-1} \sqrt{\int_{\mathbb{R}^1} g(x) n_t^0(dx)},$$

where $g(x) = \sum_{k=1}^\infty \frac{\Delta_k}{a_k} \sin^2 \frac{\sqrt{a_k}x}{2}$. For any $0 < \gamma < 1 - \frac{2}{\alpha}$ we have:

$$g(x)/|x|^{2\gamma} \le \text{const} \sum_{k=1}^\infty \frac{\Delta_k}{a_k^{1-\gamma}}.$$

Further,

$$\frac{\Delta_k}{a_k^{1-\gamma}} = \left(\frac{k^\alpha}{a_k}\right)^{1-\gamma} \cdot \frac{\ln^2 k}{k^{\alpha(1-\gamma)-1}} \le \text{const} \cdot \frac{\ln^2 k}{k^{\alpha(1-\gamma)-1}} \le \text{const} \cdot k^{-\delta},$$

where $\delta = \alpha(1-\gamma)/2 > 1$. Thus the series $\sum_{k=1}^\infty \frac{\Delta_k}{a_k^{1-\gamma}}$ converges and consequently, $g(x)/|x|^{2\gamma} \le \text{const}$. Therefore, as in Proposition 5.4.2, we deduce that for any $\beta > 1 + \frac{2}{\alpha}$

$$\|\nabla_B \mathcal{J}_\beta\|_{\bar{L}_1} < +\infty.$$

In particular, if the condition

$$(\forall \alpha > 1) \quad \lim_{k \to \infty} \frac{a_k}{k^\alpha} = \infty \qquad (N_\infty)$$

is satisfied, then for any $\beta > 1$

$$||\nabla_B \mathcal{J}_\beta||_{\tilde{L}_1} < +\infty.$$

5.4.2

We defined Bessel potential $\mathcal{J}_\beta[f]$ of function f by the equality

$$\mathcal{J}_\beta[f](x) = \int_{\mathbb{T}^\infty} f(x - y)\mathcal{J}_\beta(dy).$$

The linear operator $\mathcal{J}_\beta : \mathbb{C}(\mathbb{T}^\infty) \to \mathbb{C}(\mathbb{T}^\infty)$ is bounded and has norm equal to one. It is easy to show (for example, with the help of the Fourier transform) that $Ker\,\mathcal{J}_\beta = \{0\}$. Therefore the inverse operator $\mathcal{B}^\beta := \mathcal{J}_\beta^{-1}$ is correctly defined. We note in passing that $\mathcal{B}^2 = I - \mathcal{L}$, and for $\beta < 2$ the operator $\mathcal{B}^\beta$ coincides with the fractional degree (see [109: p. 357]) of operator $I - \mathcal{L}$, i.e., $\mathcal{B}^\beta = (I - \mathcal{L})^{\beta/2}$.

For any $\beta \geq 0$ we define a Banach space $\mathbb{C}_\beta(\mathbb{T}^\infty)$ in the following manner:

$$\mathbb{C}_0(\mathbb{T}^\infty) := \mathbb{C}(\mathbb{T}^\infty),\ \ ||f||_0 := ||f||_{\mathbb{C}(\mathbb{T}^\infty)},$$

$$(\forall \beta > 0)\quad \mathbb{C}_\beta(\mathbb{T}^\infty) := \mathrm{dom}\,\mathcal{B}^\beta,\ \ ||f||_\beta := ||\mathcal{B}^\beta f||_{\mathbb{C}(\mathbb{T}^\infty)}.$$

From the fact that the family of operators $\{\mathcal{J}_\beta\}_{\beta>0}$ forms a strongly continuous semigroup of contractions in Banach space $\mathbb{C}(\mathbb{T}^\infty)$, the following properties can easily be deduced:
1) $\mathbb{C}_\alpha(\mathbb{T}^\infty) \subset \mathbb{C}_\beta(\mathbb{T}^\infty)$ $(\alpha > \beta)$,
2) $\mathcal{B}^{\alpha+\beta} f = \mathcal{B}^\alpha \mathcal{B}^\beta f$ $(\forall f \in \mathbb{C}_{\alpha+\beta}(\mathbb{T}^\infty))$,
3) the uniform closure of $\cup_{\alpha>0}\mathbb{C}_\alpha(\mathbb{T}^\infty)$ coincides with $\mathbb{C}(\mathbb{T}^\infty)$,
4) $\mathcal{D} \subset \mathbb{C}_\alpha(\mathbb{T}^\infty)$, $\mathcal{B}^\alpha(\mathcal{D}) = \mathcal{D}$ $(\forall \alpha > 0)$.

In the following propositions, as in Sec. 5.3.7, we denote by $\partial_x u$, $\partial^2_{x,y} u$, $\ldots$, $(x, y, \ldots \in \mathbb{R}^\infty)$ the derivatives of function u along direction

$$H_A = \left\{ x = (x_i)_1^\infty \in \mathbb{R}^\infty : |x|_A^2 = \sum_{k=1}^\infty \frac{x_k^2}{a_k} < \infty \right\}.$$

Proposition 5.4.3. *Let condition (N_α) be satisfied, $\alpha > 1$. Then for any function $u \in \mathcal{D}$, any $x, y \in H_A$, and any $\gamma \geq 0$ and $\delta > \gamma + \frac{2}{\alpha}$ the following inequality is satisfied:*

$$||\partial^2_{x,y} u||_\gamma \leq \mathrm{const}\,|x|_A |y|_A\,(||\mathcal{L}u||_\delta + ||u||_\delta).$$

Proof. For $\beta > 1 + \frac{1}{\alpha}$ and for a function $u \in \mathcal{D}$, we write the identity $u = \mathcal{J}_\beta * \mathcal{B}^\beta u$. Differentiating under the integral sign, we obtain for any $\gamma \geq 0$ and $x \in H_A$

$$\mathcal{B}^\gamma \partial_x u = \partial_x \mathcal{J}_\beta * \mathcal{B}^{\beta+\gamma} u.$$

From this we get:

$$||\partial_x u||_\gamma \leq ||\partial_x \mathcal{G}_\beta||_{L_1} ||u||_{\beta+\gamma} \leq$$
$$\leq |x|_A ||\nabla_A \mathcal{G}_\beta||_{\tilde{L}_1} ||u||_{\beta+\gamma} = \text{const}\, |x|_A ||u||_{\beta+\gamma},$$

where by Proposition 5.4.2 we have $\text{const} := ||\nabla_A \mathcal{G}_\beta||_{\tilde{L}_1} < \infty$. Now applying the obtained inequality to the function $u := \partial_y u$, we arrive at the inequality

$$||\partial^2_{x,y} u||_\gamma \leq \text{const}\, |x|_A ||\partial_y u||_{\beta+\gamma} \leq \text{const}\, |x|_A |y|_A ||u||_{2\beta+\gamma}.$$

Considering that $||u||_{2\beta+\gamma} = ||\mathcal{B}^2 u||_{2\beta-2+\gamma}$, $\mathcal{B}^2 = I - \mathcal{L}$, we obtain

$$||u||_{2\beta+\gamma} \leq ||\mathcal{L}u||_{2\beta-2+\gamma} + ||u||_{2\beta-2+\gamma}.$$

Now it remains to set $\delta := 2\beta - 2 + \gamma > \gamma + \frac{2}{\alpha}$ and the required inequality is established. $\qquad\square$

Remark 5.4.3. Let condition (N_1) be satisfied. Then, in correspondence with Remark 5.4.1, we obtain the inequality

$$||\partial^2_{x,y} u||_\gamma \leq \text{const}\, |x|_B |y|_B (||\mathcal{L}u||_\delta + ||u||_\delta)$$

in which $B = (b_k)_1^\infty$ is such that $\sum_{k=1}^\infty \frac{b_k}{a_k} < \infty$, $x, y \in H_B$, and γ and δ are any numbers related by the inequality $\delta > \gamma \geq 0$.

In particular, for any $i, j \geq 1$, we shall have:

$$||\partial_i \partial_j u||_\gamma \leq \text{const}(||\mathcal{L}u||_\delta + ||u||_\delta).$$

Remark 5.4.4. Let condition (N_α) be satisfied, $\alpha > 2$. Then in correspondence with Remark 5.4.2, we obtain the inequality

$$||\partial^2_{x,y} u||_\gamma \leq \text{const}\, |x|_B |y|_B (||\mathcal{L}u||_\delta + ||u||_\delta),$$

in which $b_k = a_k \Delta_k$, $\Delta_k = k \ln^2 k$ $(k \geq 2)$, $\Delta_1 = 1$, $\gamma \geq 0$, $\delta > \gamma + \frac{4}{\alpha}$.

In particular, if condition (N_∞) is satisfied, then the mentioned inequality holds for any numbers $\gamma \geq 0$ and $\delta > \gamma$.

Theorem 5.4.1. *Let condition (N_∞) be satisfied. Then for any $\delta > 0$ and any function $f \in \mathbb{C}_\delta(\mathbb{T}^\infty)$ the function $u = \mathcal{G}_2[f]$ is a strong solution of equation $\mathcal{L}u - u = -f$, i.e. function u is twice continuously differentiable (along directions from H_A),*

$$\sum_{k=1}^\infty a_k \partial^2_k u - u = -f,$$

where the series from the left converges uniformly (and even in any $\mathbb{C}_\gamma(\mathbb{T}^\infty)$ for all $0 \leq \gamma < \delta$).

Proof. We choose a sequence of functions $\varphi_n \in \mathcal{D}$ weakly converging to the Dirac measure, and set $f_n = f * \varphi_n$, $u_n = \mathcal{G}_2[f_n]$. The sequences $\{f_n\}_1^\infty$ and $\{u_n\}_1^\infty$ are in $\mathcal{D}$ and converge in $\mathbb{C}_\delta(\mathbb{T}^\infty)$ to functions f and u respectively. Applying Proposition 5.4.3, we conclude that the sequences $\{\partial_x u_n\}_1^\infty$ and $\{\partial^2_{x,y} u_n\}_1^\infty$ are Cauchy sequences in $\mathbb{C}_\gamma(\mathbb{T}^\infty)$. Therefore in the standard manner we deduce the continuous differentiability of function u up to second order inclusive. Now we apply Remark 5.4.4 for the proof of the convergence of series $\sum_{k=1}^\infty a_k \partial^2_k u$ in $\mathbb{C}_\gamma(\mathbb{T}^\infty)$. We have

$$\left\| \sum_{k=m}^n a_k \partial^2_k u \right\|_\gamma = \left\| \sum_{k=m}^n \frac{1}{\Delta_k} b_k \partial^2_k u \right\|_\gamma \leq$$

$$\leq \max_{k \geq 1} \| b_k \partial^2_k u \|_\gamma \sum_{k=m}^n \frac{1}{\Delta_k}.$$

We set $x = (0, \ldots, \sqrt{b_k}, \ldots)$. Then $|x|_B = 1$, and consequently, for all $k = 1, 2, \ldots$

$$\|b_k \partial^2_k u\|_\gamma = \|\partial^2_{x,x} u\|_\gamma \leq \text{const} \,\|f\|_\delta.$$

Now it remains to note that the series $\sum_{k=1}^\infty 1/\Delta_k$ converges. $\qquad\square$

Remark 5.4.5. The assertions of the theorem remain true if condition (N_α), $\alpha > 2$, is satisfied. But in this connection the numbers γ and δ must be related by the inequality $\delta > \gamma + \frac{4}{\alpha}$.

If condition (N_α) is satisfied for $1 \leq \alpha \leq 2$, then in correspondence with Remark 5.4.3, the function $u = \mathcal{J}_2[f]$ is twice continuously differentiable with respect to a finite collection of variables, where the series $\sum_{k=1}^\infty a_k \partial^2_k u$ converges in L_p for any $1 < p < \infty$.

Remark 5.4.6. It would be interesting (and seems to be possible) to prove Theorem 5.4.1 for $f \in \text{Lip}_\delta(\mathbb{T}^\infty)$, $0 < \delta < 1$, where

$$\text{Lip}_\delta(\mathbb{T}^\infty) := \left\{ f \in \mathbb{C}(\mathbb{T}^\infty) : \sup_{x,y \in \mathbb{T}^\infty} \frac{|f(x) - f(y)|}{\|x - y\|^\delta} < \infty \right\}$$

and where $\|x - y\|$ is an intrinsic metric on the compact group $\mathbb{T}^\infty$ generated by elliptic operator $\mathcal{L} = \sum_{k=1}^\infty a_k \partial^2_k$ (see Chapter 7, Sec. 3), i.e.

$$\|x - y\| := \sup \left\{ \psi(x) - \psi(y); \ \psi \in D, \ \sup_{x \in \mathbb{T}^\infty} |\nabla_A \psi(x)| \leq 1 \right\}.$$

This metric can be computed explicitly:

$$\|x - y\|^2 = \sum_{k=1}^\infty \frac{1}{a_k} \|x_k - y_k\|_*^2,$$

where $\|x_k - y_k\|_*$ is the geodesic distance on the one-dimensional torus.

Chapter 6
Special classes of harmonic functions and potentials

The present chapter consists of two parts. In the first part (Sec. 6.1) we introduce the mixed norm on the set of martingales with respect to a special family of σ-algebras. Such spaces of martingales (denoted by $\mathcal{M}_{\vec{p}}$) naturally arise during tranference of the definition of a mixed norm from finite-dimensional structures (see [34]) to infinite-dimensional ones (see [17]). Many properties of spaces of martingales $\mathcal{M}_{\vec{p}}$ are analogous to properties of ordinary spaces L_p of integrable functions, where these spaces are identified if $\vec{p} = (p, p, \ldots)$ and $p > 1$. In addition, as shown in the second part (Secs. 6.2 and 6.3), the scale of spaces $\{\mathcal{M}_{\vec{p}}\}$ turns out to be more suitable to some problems of potential theory on group $\mathbb{T}^{\infty}$ than the scale of ordinary spaces $\{L_p\}$. Thus, in Sec. 6.2, considering harmonic functions in semispace $\mathbb{T}^{\infty}_{+} = \,]0, \infty[\, \times \mathbb{T}^{\infty} \subset \mathbb{R}^1 \times \mathbb{T}^{\infty}$, we apply the $\mathcal{M}_{\vec{p}}$-norm for investigation of the boundary behavior of these functions, as well as for the problem of representing them in the form of a Poisson integral. Classical finite-dimensional results on this problem are outlined in monograph [102: Chapter 2].

In Sec. 6.3 we give the infinite-dimensional analogue of the well-known Sobolev inequality (see [101: p. 141]):

$$||\rho^{\alpha-n} * f||_{L_q(\mathbb{R}^n)} \le c||f||_{L_p(\mathbb{R}^n)}, \tag{6.1}$$

where $1/q = 1/p - \alpha/n$, $0 < \alpha < n$, and ρ is the Euclidean distance in $\mathbb{R}^n$.

Our approach to the generalization of inequality (6.1) consists, first, in the fact that instead of group $\mathbb{R}^n$ we consider the infinite-dimensional abelian group $\mathbb{T}^{\infty}$. We shall not fix any metric on this group; therefore the role of function $\rho^{\alpha-n}$ is played by the kernel $\mathcal{I}_{\alpha}$ of Bessel potentials with respect to Gaussian semigroup (μ_t):

$$\mathcal{I}_{\alpha} = \frac{1}{\Gamma(\alpha/2)} \int_0^{\infty} e^{-t} t^{\alpha/2-1} \mu_t \, dt,$$

$$\widehat{\mu}_t(\theta) = \exp(-t\Psi(\theta)), \quad \Psi(\theta) = \sum_{k=1}^{\infty} a_k \theta_k^2.$$

We note that in the finite-dimensional case (i.e. in the case of group $\mathbb{R}^n$), $\mathcal{I}_{\alpha} \sim \rho^{\alpha-n}$, for $\rho \downarrow 0$.

The next step of our constructions is to pass from ordinary spaces L_p to spaces $\mathcal{M}_{\vec{p}}$ and in terms of these spaces we establish the necessary generalization of inequality (6.1):

$$\|\mathcal{J}_\alpha * f\|_{\mathcal{M}_{\vec{q}}} \leq c\|f\|_{\mathcal{M}_{\vec{p}}}, \tag{6.2}$$

where the vectors $\vec{q} = (q_k)_1^\infty$ and $\vec{p} = (p_k)_1^\infty$ are such that for any $k \geq 1$ we have $\frac{1}{q_k} = \frac{1}{p_k} - \frac{1}{\Delta_k}$, where $\sum_{k=1}^\infty \frac{1}{\Delta_k} < \alpha$.

We note that Sobolev's inequality in mixed norms on group $\mathbb{R}^n$ has been well-known for a rather long time and is given, for example, in [34]. However, the method of proof of this inequality gives a constant $c = c(n)$ which increases exponentially for $n \uparrow \infty$. This obstacle does not permit us to obtain the infinite-dimensional result from the finite-dimensional using the periodization method (see [102: chapter 7]) and the limit passage procedure for $n \uparrow \infty$.

From inequality (6.2), as in the finite-dimensional case [101: Chapter 5, Sec. 2.2], we deduce the inclusion of the space of Bessel potentials $L_{\vec{p}}^\alpha = \{\mathcal{J}_\alpha * \varphi : \varphi \in L_{\vec{p}}\}$ in the space of continuous functions $\mathbb{C}(\mathbb{T}^\infty)$.

6.1 Spaces $\mathcal{M}_{\vec{p}}$ of martingales with mixed norm

Let $(\Omega_k, \mathcal{A}_k, Q_k)_1^\infty$ be a sequence of probability spaces. We introduce the following notation:

$$\Omega_m^n := \prod_{k=m+1}^n \Omega_k, \quad \mathcal{A}_m^n := \prod_{k=m+1}^n \mathcal{A}_k, \quad P_m^n := \otimes_{k=m+1}^n Q_k,$$

$$\Omega^n := \Omega_0^n, \quad \mathcal{A}^n := \mathcal{A}_0^n, \quad P^n := P_0^n,$$

$$\Omega := \Omega_0^\infty, \quad \mathcal{A} := \mathcal{A}_0^\infty, \quad P := P_0^\infty, \quad P_m := P_m^\infty.$$

If ω_k is an elementary event in Ω_k, then $\omega_m^n := (\omega_{m+1}, \ldots, \omega_n)$ is an elementary event in Ω_m^n; in this connection we set $\omega^n := \omega_0^n$, $\omega_m := \omega_m^\infty$, $\omega := \omega_0^\infty$.

In the following, the letters f, g, and h shall denote random variables (r.v.) on $(\Omega, \mathcal{A}, P)$. We shall call random variables that depend on finitely many coordinates *cylindrical*.

Let $\mathcal{B}_n$ be the σ-algebra on Ω consisting of cylindrical events of the type $A^n \times \Omega_n^\infty$, where $A^n \in \mathcal{A}^n$. It is clear that $(\mathcal{B}_n)_1^\infty$ is an increasing family of σ-algebras and $\mathcal{B}_\infty := \vee_{n=1}^\infty \mathcal{B}_n = \mathcal{A}$. The following fact is well known (see, for example [83]):

Proposition 6.1.1. *Let $f \in L_p(\Omega, \mathcal{A}, P)$, $1 \leq p \leq \infty$. Then the martingale $(f_n, \mathcal{B}_n, P)$, where $f_n = E[f|\mathcal{B}_n]$, satisfies the following properties:*
1) $\|f\|_p = \sup_n \|f_n\|_p = \lim_{n\to\infty} \uparrow \|f_n\|_p$;
2) $\|f - f_n\|_p \to 0$ for $n \to \infty$ $(1 \leq p < \infty)$;

3) $f_n(\omega) = \int_{\Omega_n^\infty} f(\omega^n, \omega_n)\, dP_n(\omega_n)$ P-almost sure (and therefore $f_n \in L_p(\Omega^n, \mathcal{A}^n, P^n)$).

Properties 1)–3) serve for us as a starting point for the definition of a space with mixed norm on $(\Omega, \mathcal{A}, P)$. In fact, since r.v. f_n depends on a finite number of variables, then the p-norm of f_n can be replaced by a mixed norm in the sense of [28], and then a mixed norm of r.v. f is defined as the supremum of the mixed norms of f_n.

We note that spaces with mixed norms on finite Cartesian products of probability spaces (and in particular on $\mathbb{R}^n$) are an intensively developing area of study (see, for example, [34]). In the present section we introduce and investigate spaces with mixed norm on the $(\Omega, \mathcal{A}, P)$-infinite Cartesian product of probability spaces (see [17]). In this connection, the martingale approach lies at the basis of our construction.

6.1.1

We denote by $\vec{p} = (p_1, \ldots, p_n, \ldots)$ a vector with infinitely many coordinates, where $1 \le p_k \le \infty$ $(k = 1, 2, \ldots)$. We shall write the latter condition thus: $1 \le \vec{p} \le \infty$ (here and in the following, relations between coordinates of vectors are replaced by the corresponding relations between the vectors themselves). We set $\vec{p}^{\,n} := (p_1, \ldots, p_n)$, $\vec{p}_n := (p_{n+1}, \ldots)$; thus, $\vec{p} := (\vec{p}^{\,n}, \vec{p}_n)$.

The set of r.v. f on $(\Omega^n, \mathcal{A}^n, P^n)$ satisfying the inequality

$$\|f\|_{\vec{p}_n} := \|\ldots\|\,\|f(\omega_1, \ldots, \omega_{n-1}, \omega_n)\|_{p_n,\omega_n}\|_{p_{n-1},\omega_{n-1}} \cdots \|_{p_1,\omega_1} < \infty, \qquad (6.1.1)$$

where:

$$\|g(\omega_1, \ldots, \omega_k)\|_{p_k,\omega_k} := \begin{cases} \left(\int_{\Omega_k} |g(\omega_1, \ldots, \omega_k)|^{p_k}\, dQ_k(\omega_k)\right)^{1/p_k}, & p_k < \infty \\ \operatorname*{ess\,sup}_{\omega_k \in \Omega_k} |g(\omega_1, \ldots, \omega_k)|, & p_k = \infty \end{cases}$$

is called a space with mixed norm on $(\Omega^n, \mathcal{A}^n, P^n)$ and is denoted by $L_{\vec{p}^{\,n}}$.

This definition is different from the generally accepted ones (see [34], [28]) only by the order of taking the norm in formula (6.1.1). This form is convenient for transference to the infinite-dimensional case.

The set of r.v. $f \in L_1(\Omega, \mathcal{A}, P)$ such that

$$\|f\|_{\vec{p}} := \sup_n \|f_n\|_{\vec{p}^{\,n}} < \infty, \qquad (6.1.2)$$

where

$$f_n := f \cdot P_n = \int_{\Omega_n^\infty} f(\omega^n, \omega_n)\, dP_n(\omega_n), \qquad (6.1.3)$$

shall be called the space with mixed norm on $(\Omega, \mathcal{A}, P)$ and denoted by $L_{\vec{p}}$ (see [17]).

We note that in fact $L_{\vec{p}}$, as usual, means the aggregate of classes of r.v. that coincide almost surely (a.s.). In addition, $L_{\vec{p}} \subset L_{\vec{q}} \subset L_1$, if $\vec{p} \geq \vec{q}$.

From (6.1.2) and from the fact that $\| \cdot \|_{\vec{p}^n}$ is a norm, it follows that $\| \cdot \|_{\vec{p}}$ is a norm.

Proposition 6.1.2. *In the notation of (6.1.3) the following inequality holds:*

$$\|f_n\|_{\vec{p}^n} \leq \|f_{n+1}\|_{\vec{p}^{n+1}}.$$

Proof. Set $g = f_{n+1}$. Using the martingale property, we get that $f_n = \int_{\Omega_{n+1}} g \, dQ_{n+1}$. From formula (6.1.1) it follows that for the proof of the desired inequality it is sufficient to verify that P^n-a.s. with respect to $\omega^n \in \Omega^n$, the following inequality holds:

$$\left| \int_{\Omega_{n+1}} g(\omega^n, \omega_{n+1}) \, dQ_{n+1}(\omega_{n+1}) \right| \leq \|g(\omega^n, \omega_{n+1})\|_{p_{n+1}, \omega_{n+1}}.$$

But the latter inequality is obvious because $p_{n+1} \geq 1$. $\qquad\qquad\square$

Corollary 6.1.1. *The following equality holds:*

$$\|f\|_{\vec{p}} = \lim_{n \to \infty} \|f_n\|_{\vec{p}^n} = \lim_{n \to \infty} \uparrow \|f_n\|_{\vec{p}}.$$

Corollary 6.1.2. *If $p_k = p$ $(k = 1, 2, \ldots)$, then $L_{\vec{p}} = L_p(\Omega, \mathcal{A}, P)$.*

The truth of this corollary follows from part 1) of Proposition 6.1.1 and the Fubini theorem. $\qquad\qquad\square$

Proposition 6.1.3. *If we have $p_k = p$ for $k > n$, then*

$$\|f\|_{\vec{p}} = \| \|f(\omega^n, \omega_n)\|_{p,\omega_n} \|_{\vec{p}^n, \omega^n}.$$

Proof. By Corollary 6.1.2, P^n-a.s. with respect to $\omega^n \in \Omega^n$, the following equality is satisfied:

$$\|f(\omega^n, \omega_n)\|_{p,\omega_n} = \|f(\omega^n, \omega_n)\|_{\vec{p}_n, \omega_n}.$$

But it is obvious that $\|f\|_{\vec{p}} = \| \|f(\omega^n, \omega_n)\|_{\vec{p}_n, \omega_n} \|_{\vec{p}^n, \omega^n}.$ $\qquad\square$

Proposition 6.1.4. *Space $L_{\vec{p}}$ is a normed ideal space (see [70: p. 139]), i.e. if $f \in L_{\vec{p}}$, and $|g| \leq |f|$, then $\|g\|_{\vec{p}} \leq \|f\|_{\vec{p}}$.*

Proof. We shall verify the relation

$$\| |f| \|_{\vec{p}} = \|f\|_{\vec{p}}, \tag{6.1.4}$$

which implies the truth of the given proposition. We have $\||f|\|_{\vec{p}} = \sup_n \||f| \cdot P_n\|_{\vec{p}^n}$. Applying Proposition 6.1.3, we obtain

$$\||f| \cdot P_n\|_{\vec{p}^n} = \|\|f(\omega^n, \omega_n)\|_{\vec{1}, \omega_n}\|_{\vec{p}^n, \omega^n} = \|f\|_{\vec{q}},$$

where $\vec{q} = (p_1, \ldots, p_n, 1, 1, \ldots)$. Since $\vec{q} \le \vec{p}$, then $\|f\|_{\vec{q}} \le \|f\|_{\vec{p}}$. From this it follows that $\||f|\|_{\vec{p}} \le \|f\|_{\vec{p}}$. The reverse inequality is obvious. $\qquad\square$

Remark 6.1.1. By formula (6.1.4) it is natural to assume that if $f \notin L_{\vec{p}}$, then $\|f\|_{\vec{p}} := \||f|\|_{\vec{p}} = \infty$.

Theorem 6.1.1. *Let* $1/\vec{p} + 1/\vec{q} = 1$. *Then for arbitrary r.v.* f *the following equality holds:*

$$\|f\|_{\vec{p}} = \sup_{g \in M_{\vec{q}}} \int_\Omega |f \cdot g|\, dP = \sup_{g \in M_{\vec{q}}} \int_\Omega f \cdot g\, dP, \qquad (6.1.5)$$

where $M_{\vec{q}}$ *is the set of finite-valued cylindrical r.v.* g *such that* $\|g\|_{\vec{q}} = 1$..

The proof of the theorem follows from the following two lemmas.

Lemma 6.1.1. *(Hölder's inequality) For arbitrary r.v.* f *and* g *the following inequality holds:*

$$\int_\Omega |f \cdot g|\, dP \le \|f\|_{\vec{p}} \|g\|_{\vec{q}}, \quad 1/\vec{p} + 1/\vec{q} = 1.$$

The proof is obtained by an application of Fatou's lemma to the finite-dimensional Hölder inequality for mixed norms (see [34], [28]).

Lemma 6.1.2. *(inverse Hölder inequality) Let* $1/\vec{p} + 1/\vec{q} = 1$ *and r.v.* f *be such that for any finite-valued cylindrical r.v.* g *the following inequality is satisfied:*

$$\int_\Omega f \cdot g\, dP \le c\|g\|_{\vec{q}},$$

then $\|f\|_{\vec{p}} \le c$.

Proof. Let g be a finite-valued r.v. depending only on the first n coordinates. We have:

$$\int_\Omega (f \cdot P_n) g\, dP = \int_\Omega f(g \cdot P_n)\, dP = \int_\Omega f \cdot g\, dP \le c\|g\|_{\vec{q}^n}.$$

Applying the well-known finite-dimensional result (see [28: § 3, Lemma 1]), we get $\|f \cdot P_n\|_{\vec{p}_n} \le c$. Consequently, $\|f\|_{\vec{p}} \le c$. $\qquad\square$

Remark 6.1.2. Equality (6.1.5) can be interpreted in the following manner: the associated space $L_{\vec{p}}^\times$ (see [70: Chapter 6, § 2]) coincides with space $L_{\vec{q}}$. Therefore $L_{\vec{p}}^{\times\times} = L_{\vec{p}}$.

Corollary 6.1.3. *Space $L_{\vec{p}}$ is complete and the following property is satisfied in it: if a sequence of r.v. (f_n) is such that $0 \leq f_n \uparrow f$ a.s. and $f \in L_{\vec{p}}$, then $||f_n||_{\vec{p}} \uparrow ||f||_{\vec{p}}$.*

The proof follows from Remark 6.1.2 and from [70: Chapter 4, § 3, Theorem 7 and Chapter 6, § 1, Theorem 4].

Corollary 6.1.4. *The Fatou property is satisfied in space $L_{\vec{p}}$: if $f_n, f \in L_{\vec{p}}$ and $f_n \to f$ on probability, then*

$$||f||_{\vec{p}} \leq \liminf_{n \to \infty} ||f_n||_{\vec{p}}.$$

The proof follows from Corollary 6.1.3 and from [70: Chapter IV, § 3, Lemma 4].

It is also not difficult to verify the following statement of the Fatou property: if $f_n \geq 0$ a.s. $(n = 1, 2, \ldots)$, then

$$|| \liminf_{n \to \infty} f_n||_{\vec{p}} \leq \liminf_{n \to \infty} ||f_n||_{\vec{p}}. \tag{6.1.6}$$

Proposition 6.1.5. *(integral inequality) Let $\varphi \in L_1(\Omega \times X)$, where X is a space with σ-finite measure μ. Then*

$$\left|\left| \int_X \varphi(\omega, x)\, d\mu(x) \right|\right|_{\vec{p}, \omega} \leq \int_X ||\varphi(\omega, x)||_{\vec{p}, \omega} d\mu(x).$$

The proof is obtained by an application of Remark 6.1.2 to the well-known Minkowski inequality for ideal spaces (see [71: pp. 65–66]). $\qquad\square$

Proposition 6.1.6. *(Chebyshev inequality) Let f be an r.v. and $1_{\{|f| \geq N\}}$ the indicator of event $\{|f| \geq N\}$. The following inequality holds:*

$$||1_{\{|f| \geq N\}}||_{\vec{p}} \leq ||f||_{\vec{p}}/N.$$

Proof. Let $f \in L_{\vec{p}}$. Then r.v. $|f| \cdot P_n$ depends only on $\omega_1, \ldots, \omega_n$. Applying the usual Chebyshev inequality in variable ω_n to it, we get

$$||1_{\{|f| \cdot P_n \geq N - \varepsilon\}}||_{\vec{p}} \leq |||f| \cdot P_n||_{\vec{p}}/N - \varepsilon.$$

Since $|f| \cdot P_n \to |f|$ a.s. for $n \to \infty$, it is easy to see that

$$1_{\{|f| \geq N\}} \leq \liminf_{n \to \infty} 1_{\{|f| \cdot P_n \geq N - \varepsilon\}}.$$

Therefore, applying the Fatou property (4.1.6), we have

$$||1_{\{|f| \geq N\}}||_{\vec{p}} \leq || \liminf_{n \to \infty} 1_{\{|f| \cdot P_n \geq N - \varepsilon\}}||_{\vec{p}} \leq$$

$$\leq \liminf_{n \to \infty} ||1_{\{|f| \cdot P_n \geq N - \varepsilon\}}||_{\vec{p}} \leq \liminf_{n \to \infty} |||f| \cdot P_n||_{\vec{p}}/(N - \varepsilon) =$$

$$= ||f||_{\vec{p}}/(N - \varepsilon).$$

Letting ε go to zero, we obtain the required result. $\qquad\square$

6.1.2

We shall consider four groups of convergence properties:

Group (I)

C.1) If sequence $(f_k) \subset L_{\vec{p}}$ is such that $f_k \downarrow 0$ a.s., then $||f_k||_{\vec{p}} \downarrow 0$ (see [70: Chapter IV, § 3.2]).

C.2) The norm of space $L_{\vec{p}}$ is uniformly absolutely continuous, i.e. if $f \in L_{\vec{p}}$, then $\forall \varepsilon > 0 \; \exists \delta > 0$:

$$\forall A \in \mathcal{A} : P(A) < \delta \Rightarrow \; ||f \cdot 1_A||_{\vec{p}} < \varepsilon.$$

C.3) The norm of space $L_{\vec{p}}$ is absolutely continuous, i.e. if $f \in L_{\vec{p}}$ and the sequence of events $(A_k) \subset \mathcal{A}$ is such that $P(A_k) \to 0$, then $||f \cdot 1_{A_k}||_{\vec{p}} \to 0$, $k \to \infty$.

C.4) The dominated convergence property is satisfied, i.e. if the sequence $(f_k) \subset L_{\vec{p}}$ converges on probability to f and $|f_k| \le g \in L_{\vec{p}}$ ($k = 1, 2, \ldots$), then $f \in L_{\vec{p}}$ and $||f - f_k||_{\vec{p}} \to 0$.

C.5) The Lévy property is satisfied, i.e. if $(f_k) \subset L_{\vec{p}}$, $f_k \uparrow f$ a.s. and $\sup_k ||f_k||_{\vec{p}} < \infty$, then $f \in L_{\vec{p}}$, $||f - f_k||_{\vec{p}} \downarrow 0$.

Group (II)

C.6) For any r.v. $f \in L_{\vec{p}}$ we have $||f \cdot 1_{\{|f|>N\}}||_{\vec{p}} \downarrow 0$ for $N \uparrow \infty$.

C.7) L_∞ is dense in $L_{\vec{p}}$.

Group (III)

C.8) For any r.v. $f \in L_{\vec{p}}$ we have $f \cdot P_n \to f$ in $L_{\vec{p}}$.

C.9) $\cup_{n=1}^\infty L_{\vec{p}^n}$ is dense in $L_{\vec{p}}$.

Group (IV)

C.10) The set of bounded cylindrical r.v. is dense in $L_{\vec{p}}$.

C.11) The set of finite-valued cylindrical r.v. is dense in $L_{\vec{p}}$.

Proposition 6.1.7. *Within each of the above-introduced four groups, the properties are equivalent* ($1 \le p \le \infty$).

Proof. We shall show the equivalence of the properties of Group (I).

C.1) $\Rightarrow$ C.2). We assume that this is not so: let there exist $f \in L_{\vec{p}}$, $\varepsilon > 0$ and a sequence of events $A_k \in \mathcal{A}_k$ with $P(A_k) < 1/2^k$ ($k = 1, 2, \ldots$) such that $||f 1_{A_k}||_{\vec{p}} \ge \varepsilon$. We denote $B_n := \cup_{k=n}^\infty A_k$. It is obvious that the B_n's decrease monotonically and $P(B_n) < 1/2^{n-1}$. Therefore $|f| 1_{B_n} \downarrow 0$ a.s. for $n \uparrow \infty$. However, $||f \cdot 1_{B_n}||_{\vec{p}} \ge ||f \cdot 1_{A_n}||_{\vec{p}} \ge \varepsilon$. Contradiction!

C.2) $\Rightarrow$ C.3) Trivial.

C.3) $\Rightarrow$ C.4) The fact that $f \in L_{\vec{p}}$ follows from Corollary 6.1.4. The convergence $f_k \to f$ in $L_{\vec{p}}$ follows from properties of ideal spaces of functions (see [110], [70]).

C.4) $\Rightarrow$ C.5) The fact that $f \in L_{\vec{p}}$ again follows from Corollary 6.1.4. It remains to note that $(f - f_k) \downarrow 0$ a.s.

C.5) $\Rightarrow$ C.1) Let $(f_k) \subset L_{\vec{p}}$, $f_k \downarrow 0$ a.s. Denoting $g_k = f_1 - f_k$, we get that $g_k \uparrow f_1$ a.s. Therefore $\|f_1 - g_k\|_{\vec{p}} \downarrow 0$, i.e., $\|f_k\|_{\vec{p}} \downarrow 0$.

We shall now prove the equivalence of properties C.6) and C.7) of Group (II).

C.6) $\Rightarrow$ C.7) We shall consider the bounded r.v. $g_N = f \cdot 1_{\{|f| \leq N\}}$. We have

$$\|f - g_N\|_{\vec{p}} = \|f \cdot 1_{\{|f| > N\}}\|_{\vec{p}} \to 0 \text{ for } N \to \infty.$$

C.7) $\Rightarrow$ C.6). Let r.v. g be such that $\|g\|_{\infty} \leq c$ and $\|f - g\|_{\vec{p}} < \varepsilon$. Then:

$$\|f \cdot 1_{\{|f| > N\}}\|_{\vec{p}} \leq \|(f - g)1_{\{|f| > N\}}\|_{\vec{p}} + \|g \cdot 1_{\{|f| > N\}}\|_{\vec{p}} \leq$$
$$\leq \|f - g\|_{\vec{p}} + c\|1_{\{|f| > N\}}\|_{\vec{p}}.$$

Using the Chebyshev inequality (Proposition 6.1.6), we choose N such that $\|1_{\{|f| \geq N\}}\|_{\vec{p}} < \varepsilon/c$.

We shall give the proof of the equivalence of the properties in Group (III).

C.8) $\Rightarrow$ C.9). Trivial.

C.9) $\Rightarrow$ C.8). We shall show that if there exists a subsequence n_k such that $\|f - f \cdot P_{n_k}\|_{\vec{p}} \geq c > 0$ $(k = 1, 2, \ldots)$, then for any cylindrical r.v. $g \in L_{\vec{p}}$ we have $\|f - g\|_{\vec{p}} \geq c/2$. Suppose not: let $g \in \cup_{n=1}^{\infty} L_{\vec{p}^n}$ exist such that $\|f - g\|_{\vec{p}} < c/2$. Then

$$\|f - g\|_{\vec{p}} = \|(f - f \cdot P_{n_k}) - (g - f \cdot P_{n_k})\|_{\vec{p}} \geq$$
$$\geq \|f - f \cdot P_{n_k}\|_{\vec{p}} - \|g - f \cdot P_{n_k}\|_{\vec{p}}.$$

For sufficiently large k we get

$$\|g - f \cdot P_{n_k}\|_{\vec{p}} = \|(g - f) \cdot P_{n_k}\|_{\vec{p}} \leq \|g - f\|_{\vec{p}} < c/2.$$

Consequently, $\|f - g\|_{\vec{p}} \geq c - c/2 = c/2$. Contradiction.

The equivalence of the properties in Group (IV) is obvious. $\square$

Proposition 6.1.8. 1) *For $1 \leq \vec{p} \leq \infty$, the following implications hold:*

$$(\text{I}) \Rightarrow (\text{IV}) \Leftrightarrow \left\{ \begin{array}{l} (\text{II}) \\ (\text{III}) \end{array} \right.$$

(in this connection any other implication between any two groups of properties does not hold).

2) *If $\sup_k p_k < \infty$, then* $(\text{I}) \Leftrightarrow (\text{II}) \Leftrightarrow (\text{III}) \Leftrightarrow (\text{IV})$.

Proof. 1) It is obviously sufficient to establish:

$$(\mathrm{I}) \Rightarrow (\mathrm{II}) + (\mathrm{III}) \Rightarrow (\mathrm{IV}).$$

Implication (I) $\Rightarrow$ (II) follows from the obvious implication C.1) $\Rightarrow$ C.6). We obtain the proof of (I) $\Rightarrow$ (III) from the estimate

$$\|f - f \cdot P_n\|_{\vec{p}} \leq \|f \cdot 1_{\{|f| \leq N\}} - (f \cdot 1_{\{|f| \leq N\}}) \cdot P_n\|_{\vec{p}} + 2\|f \cdot 1_{\{|f| > N\}}\|_{\vec{p}}.$$

And finally, we establish (IV). By (II), for r.v. $f \in L_{\vec{p}}$ we can find $g \in L_\infty$ such that $\|f - g\|_{\vec{p}} < \varepsilon/2$, and therefore by (III) we choose n large enough that $\|g - g \cdot P_n\|_{\vec{p}} < \varepsilon/2$. We get

$$\|f - g \cdot P_n\|_{\vec{p}} \leq \|f - g\|_{\vec{p}} + \|g - g \cdot P_n\|_{\vec{p}} < \varepsilon.$$

An example when (III) $\Rightarrow$ (II) is not true is given by Proposition 6.1.10. The falsity of the remaining implications is easily demonstrated for $\vec{p} = \vec{\infty}$.

2) Considering the first part of the given proposition, it is sufficient to verify the implications (III) $\Rightarrow$ (IV) and (II) $\Rightarrow$ (I).

Implication (III) $\Rightarrow$ (IV) follows from the known finite-dimensional results (see [28: § 5]). For the proof of implication (II) $\Rightarrow$ (I) we verify C.7) $\Rightarrow$ C.3). In fact, let f and (A_k) be as required by C.3), and let r.v. $g \in L_\infty$ be such that $\|f - g\|_{\vec{p}} < \varepsilon$. Denoting $r := \sup_k p_k$, we have:

$$\|f \cdot 1_{A_k}\|_{\vec{p}} \leq \|(f - g) \cdot 1_{A_k}\|_{\vec{p}} + \|g \cdot 1_{A_k}\|_{\vec{p}} \leq$$
$$\leq \|f - g\|_{\vec{p}} + \|g\|_\infty \|1_{A_k}\|_{\vec{p}} \leq \varepsilon + \|g\|_\infty \|1_{A_k}\|_r.$$

It remains to choose k large enough that $\|1_{A_k}\|_r < \varepsilon/\|g\|_\infty$. $\qquad\square$

Theorem 6.1.2. *If $\sup_k p_k < \infty$, then in space $L_{\vec{p}}$, equivalent groups of properties* (I), (II), (III), *and* (IV) *are satisfied.*

The proof is based on the following general lemma, which was brought to our attention by V. M. Kadets.

Let X be a normed lattice and $L_q(X)$ a space of random variables integrable with exponent q (in the Bochner sense) on some probability space $(W, \mathcal{B}, \mu)$ with values in X.

Lemma 6.1.3. *Let $1 \leq p, q < \infty$, and for any positive elements $a, b \in X$ let the following inequality be satisfied:*

$$\|a + b\|^p \geq \|a\|^p + \|b\|^p.$$

We denote $r = \max\{p, q\}$. Then for any two positive r.v. $x, y \in L_q(X)$, the following inequality holds:

$$\|x + y\|_{L_q(X)}^r \geq \|x\|_{L_q(X)}^r + \|y\|_{L_q(X)}^r. \tag{6.1.7}$$

Proof of the lemma. First let $p \geq q$. Applying the condition of the lemma and the inequality converse to the triangle inequality (obtained when integration is carried out with exponent less than one), we have

$$||x + y||_{L_q(X)}^p = \left(\int_W ||x(w) + y(w)||^q d\mu(w) \right)^{p/q} \geq$$

$$\geq \left(\int_W (||x(w)||^p + ||y(w)||^p)^{q/p} d\mu(w) \right)^{p/q} \geq$$

$$\geq \left(\int_W ||x(w)||^{p \cdot q/p} d\mu(w) \right)^{p/q} + \left(\int_W ||y(w)||^{p \cdot q/p} d\mu(w) \right)^{p/q} =$$

$$= ||x||_{L_q(X)}^p + ||y||_{L_q(X)}^p.$$

If $p < q$, then it is easy to see that the inequality

$$||a + b||^q \geq ||a||^q + ||b||^q$$

is also satisfied in X , and we arrive again at the first case. $\qquad\square$

Proof of Theorem 6.1.2. Let r.v. $g, h \in L_{\vec{p}}$, $g, h \geq 0$, $r = \sup_k p_k$. Considering that $L_{\vec{p}^n} = L_{p_1}(X)$, where $X := L_{p_2,\dots,p_n}$ and, applying inequality (6.1.7), we get by induction

$$||g \cdot P_n + h \cdot P_n||_{\vec{p}^n}^r \geq ||g \cdot P_n||_{\vec{p}^n}^r + ||h \cdot P_n||_{\vec{p}^n}^r.$$

Passing to the limit in this inequality for $n \to \infty$, we get

$$||g + h||_{\vec{p}}^r \geq ||g||_{\vec{p}}^r + ||h||_{\vec{p}}^r. \tag{6.1.8}$$

Now it is easy to verify C.6). Let $f \in L_{\vec{p}}$, $f \geq 0$. Setting $g = f \cdot 1_{\{f > N\}}$, $h = f \cdot 1_{\{f \leq N\}}$, from inequality (6.1.8) we obtain:

$$||f \cdot 1_{\{f > N\}}||_{\vec{p}}^r \leq ||f||_{\vec{p}}^r - ||f \cdot 1_{\{f \leq N\}}||_{\vec{p}}^r.$$

Now the proof is completed by an application of Corollary 6.1.3. $\qquad\square$

In conclusion of this section, we consider the space $L_{\vec{p}}$ under the condition that $\sup_k p_k = \infty$. For simplicity we assume that the probability spaces $(\Omega_k, \mathscr{A}_k, Q_k)$ are continuous for all $k \geq 1$.

Proposition 6.1.9. *If* $\sup_k p_k = \infty$, *then the properties of Group* (III) *are not satisfied.*

Proof. 1) Let $\limsup_{k \to \infty} p_k < \infty$. It is not difficult to see that it is sufficient to consider the case $\vec{p} = (\infty, p_2, \dots)$, where $p_k \leq r$ for $k \geq 2$. For $k \geq 2$ we choose the events $A_k \in \mathscr{A}_k$ such that $Q_k(A_k) = \frac{1}{2}$ and we define r.v.

$$g_k(\omega_k) := \begin{cases} 1, & \omega_k \in A_k, \\ -1, & \omega_k \notin A_k. \end{cases}$$

It is obvious that $||g_k||_1 = ||g_k||_{p_k} = 1$ and $g_k \cdot Q_k = 0$. We define r.v. f in the following manner:

$$f(\omega) := \sum_{k=1}^{\infty} g_1^{(k)}(\omega_1) g_2(\omega_2) \dots g_k(\omega_k),$$

where $\left\{ g_1^{(k)}(\omega_1) \right\}$ is a family of indicators with pairwise disjoint supports. It is easy to see that $||f||_{\vec{p}} = 1$. Further,

$$f \cdot P_n = \sum_{k=1}^{n} g_1^{(k)} g_2 \dots g_k + \sum_{k=n+1}^{\infty} g_1^{(k)} g_2 \dots g_n (g_{n+1} \cdot Q_{n+1}) \dots (g_k \cdot Q_k) =$$

$$= \sum_{k=1}^{n} g_1^{(k)} g_2 \dots g_k,$$

$$||f - f \cdot P_n||_{\vec{p}} = \left\| \sum_{k=n+1}^{\infty} g_1^{(k)} g_2 \dots g_k \right\|_{\vec{p}} = 1.$$

Thus, C.8) is not satisfied.

2) Now let $\limsup\limits_{k \to \infty} p_k = \infty$. Without loss of generality, we may assume that $\lim_{k \to \infty} \ln k / p_k = 0$. We denote $a_k := 1 - 1/2k^2$ and in every σ-algebra $\mathcal{A}_k$ we fix an event A_k such that $Q_k(A_k) = a_k$. We set:

$$B_n := \left(\prod_{k=1}^{n} A_k \right) \times \Omega_n^{\infty}, \quad f_n := \left(\prod_{k=1}^{n} a_k \right)^{-1} 1_{B_n}.$$

It is obvious that $(f_n, \mathcal{B}_n, P)$ is a non-negative martingale. Since $\prod_{k=1}^{\infty} a_k > 0$, this martingale is bounded. Denoting

$$f := \lim_{n \to \infty} f_n = \left(\prod_{k=1}^{\infty} a_k \right)^{-1} 1_{B_\infty},$$

we get that $f_n = f \cdot P_n$. In addition, $||f_n||_{\vec{p}} = \prod_{k=1}^{n} a_k^{1/p_k - 1}$. We have

$$f - f_n = \left(\left(\prod_{k=1}^{\infty} a_k \right)^{-1} - \left(\prod_{k=1}^{n} a_k \right)^{-1} \right) 1_{B_\infty} - \left(\prod_{k=1}^{n} a_k \right)^{-1} 1_{B_n \setminus B_\infty}.$$

We set $\widetilde{B}_n := \left(\prod_{k=1}^{n} A_k \right) \times \overline{A}_{n+1} \times \Omega_{n+1}^{\infty}$, where $\overline{A}_{n+1} = \Omega_{n+1} \setminus A_{n+1}$. It is obvious that $\widetilde{B}_n \subset B_n \setminus B_\infty$, and therefore

$$\|f - f_n\|_{\vec{p}} \geq \|(f - f_n)1_{\widetilde{B}_n}\|_{\vec{p}} = \left(\prod_{k=1}^{n} a_k\right)^{-1} \|1_{\widetilde{B}_n}\|_{\vec{p}} =$$

$$= \left(\prod_{k=1}^{n} a_k\right)^{-1} \left(\prod_{k=1}^{n} a_k^{1/p_k}\right) (1 - a_{n+1})^{1/p_{n+1}} =$$

$$= \prod_{k=1}^{n} a_k^{1/p_k - 1} \left(\frac{1}{2}\right)^{1/p_{n+1}} (n+1)^{-2/p_{n+1}} =$$

$$= \|f_n\|_{\vec{p}} \left(\frac{1}{2}\right)^{1/p_{n+1}} \exp\left(-\frac{2\ln(n+1)}{p_{n+1}}\right).$$

Consequently,

$$\liminf_{n\to\infty} \|f - f_n\|_{\vec{p}} \geq \|f\|_{\vec{p}} > 0. \qquad \square$$

From Propositions 6.1.8 and 6.1.9 it follows that if $\sup_k p_k = \infty$, then groups of properties (I) and (IV) are not satisfied. A different picture holds for group (II).

Proposition 6.1.10. *Let there be infinities among the coordinates of vector $\vec{p}$.*
1) *If in vector $\vec{p}$ after the first infinite coordinate all the remaining coordinates are also infinity, then (II) is satisfied.*
2) *If in vector $\vec{p}$ after an infinite coordinate there is a finite one, then (II) is not satisfied.*

The proof of part 1) is trivial. Part 2) can be obtained by means of modernization of an example given in [28: §5]. $\qquad \square$

6.1.3

In this section we consider other properties of spaces $L_{\vec{p}}$, such as reflexivity, uniform convexity, etc.

Theorem 6.1.3. *If $\sup_k p_k < \infty$, then $L_{\vec{p}}^* = L_{\vec{q}}$ ($1/\vec{p} + 1/\vec{q} = 1$). Otherwise this is not true.*

The proof follows from Theorems 6.1.1 and 6.1.2, Proposition 6.1.9, and from [70: Chapter VI, § 1, Corollary 1].

Corollary 6.1.5. *If $\sup_k p_k < \infty$ and $\inf_k p_k > 1$, then the space $L_{\vec{p}}$ is reflexive. Otherwise this is not true.*

Theorem 6.1.4. *If $\sup_k p_k < \infty$ and $\inf_k p_k > 1$, then the space $L_{\vec{p}}$ is uniformly convex. Otherwise, generally speaking, $L_{\vec{p}}$ is not uniformly convex.*

Proof. Let $\sup_k p_k < \infty$ and $\inf_k p_k > 1$. Analysis of Lemma 1 of [28: § 8] shows that there exists $1 < r < 2$, not depending on n, such that the following inequality holds:

$$(\|f\|_{\vec{p}}^r + \|g\|_{\vec{p}}^r)^{1/r} \cdot 2^{1/r'} \geq (\|f + g\|_{\vec{p}}^{r'} + \|f - g\|_{\vec{p}}^{r'})^{1/r'}, \qquad (6.1.9)$$

where $f, g \in L_{\vec{p}^n}$, $1/r + 1/r' = 1$. By passage to the limit we get that inequality (6.1.9) holds for any $f, g \in L_{\vec{p}}$. From (6.1.9) it follows that for two sequences (f_k) and (g_k) from $L_{\vec{p}}$ such that $\|f_k\|_{\vec{p}} \to 1$, $\|g_k\|_{\vec{p}} \to 1$ and $\|f_k + g_k\|_{\vec{p}} \to 2$, we have $\|f_k - g_k\|_{\vec{p}} \to 0$. But this means that $L_{\vec{p}}$ is uniformly convex.

We shall prove the second assertion of the theorem (for simplicity we again assume that $(\Omega_k, \mathcal{A}_k, Q_k)$ are continuous probability spaces). Without loss of generality, we may assume that either $\lim_{k\to\infty} p_k = \infty$ or $\lim_{k\to\infty} p_k = 1$. If the first equality is satisfied, then $L_{\vec{p}}$ cannot be uniformly convex, since uniform convexity implies the satisfaction of (I), and this is not true by Proposition 6.1.9.

Let $\lim_{k\to\infty} p_k = 1$ and $A_k \in \mathcal{A}_k$ be such that $Q_k(A_k) = \frac{1}{2}$. We denote $f_k(\omega_k) := 2 \cdot 1_{A_k}(\omega_k)$, $g_k(\omega_k) := 2 \cdot 1_{\overline{A}_k}(\omega_k)$. We have

$$\|f_k\|_{\vec{p}} = \|g_k\|_{\vec{p}} = \|g_k\|_{p_k} = \left(2^{p_k} \cdot \frac{1}{2}\right)^{1/p_k} \to 1,$$

but on the other hand

$$\|f_k - g_k\|_{\vec{p}} = \|f_k + g_k\|_{\vec{p}} = 2. \qquad \square$$

We shall not give the proofs of the following theorems, since they exactly copy the proofs of the corresponding facts from [28].

Theorem 6.1.5. *(Riesz property) Let* $\sup_k p_k < \infty$. *If* $f_n, f \in L_{\vec{p}}$ *are such that for* $n \to \infty$ *we have* $\|f_n\|_{\vec{p}} \to \|f\|_{\vec{p}}$ *and* $f_n \to f$ *a.s., then* $\|f - f_n\|_{\vec{p}} \to 0$.

Theorem 6.1.6. *(Radon property) Let* $\sup_k p_k < \infty$ *and* $\inf_k p_k > 1$. *If* $f_n, f \in L_{\vec{p}}$ *are such that for* $n \to \infty$ *we have* $\|f_n\|_{\vec{p}} \to \|f\|_{\vec{p}}$ *and* $f_n \to f$ *weakly, then* $\|f - f_n\|_{\vec{p}} \to 0$.

We denote the closure of L_∞ in $L_{\vec{q}}$ by $S_{\vec{q}}$.

Theorem 6.1.7. *Let* $1 \leq \vec{p} \leq \infty$ *and let sequence* $(f_n) \subset L_{\vec{p}}$ *be such that for any r.v.* $g \in S_{\vec{q}}$ $(1/\vec{p} + 1/\vec{q} = 1)$, $\lim_{n\to\infty} \int_\Omega f_n g \, dP$ *exists and is finite. Then* $\exists f \in L_{\vec{p}}$ *such that* $\forall g \in S_{\vec{q}}$ *the following relation is satisfied:*

$$\int_\Omega f \cdot g \, dP = \lim_{n\to\infty} \int_\Omega f_n \cdot g \, dP. \qquad (6.1.10)$$

Theorem 6.1.8. *Let* $1 \leq \vec{p} \leq \infty$ *and let sequence* $(f_n) \subset L_{\vec{p}}$ *be such that for any r.v.* $g \in L_{\vec{q}}$ $(1/\vec{p} + 1/\vec{q} = 1)$, $\lim_{n\to\infty} \int_\Omega f_n \cdot g \, dP$ *exists and is finite. Then* $\exists f \in L_{\vec{p}}$ *such that* $\forall g \in L_{\vec{q}}$ *the relation* (6.1.10) *is satisfied.*

We note that Theorem 6.1.8, generally speaking, does not follow from Theorem 6.1.7 (see [28: § 5]).

6.1.4

We shall denote by $\mathcal{M}_{\vec{p}}$ the space of martingales $f = (f_n, \mathcal{B}_n, P)$ on $(\Omega, \mathcal{A})$ such that

$$\|f\|_{\vec{p}} := \sup_n \|f_n\|_{\vec{p}} < \infty.$$

It can be verified without difficulty that $\mathcal{M}_{\vec{p}}$ is a Banach space, where $L_{\vec{p}}$ is identified in a natural manner with the subspace of $\mathcal{M}_{\vec{p}}$ consisting of uniformly integrable martingales.

Theorem 6.1.9. *If* $\liminf_{k \to \infty} p_k > 1$, *then* $L_{\vec{p}} = \mathcal{M}_{\vec{p}}$. *If* $\liminf_{k \to \infty} p_k = 1$, *then* $L_{\vec{p}}$ *can be isometrically imbedded in* $\mathcal{M}_{\vec{p}}$.

The truth of the theorem follows from the next two Propositions:

Proposition 6.1.11. *Let* $(f_n, \mathcal{B}_n, P) \in \mathcal{M}_{\vec{p}}$. *Then there exist two non-negative martingales* $(g_n, \mathcal{B}_n, P)$ *and* $(h_n, \mathcal{B}_n, P)$ *belonging to* $\mathcal{M}_{\vec{p}}$ *and such that* $f_n = g_n - h_n$.

Proof. Since $\sup_n \|f_n\|_1 < \infty$, it is possible to apply the Krickeberg decomposition (see [84: p. 64]). In fact, we set $g_n = \lim_{k \to \infty} \uparrow E[f_k^+ | \mathcal{B}_n]$. Then we get

$$\|g_n\|_{\vec{p}} = \lim_{k \to \infty} \|E[f_k^+ | \mathcal{B}_n]\|_{\vec{p}} \leq \sup_k \|f_k^+\|_{\vec{p}} \leq \sup_k \|f_k\|_{\vec{p}},$$

and thus $(g_n, \mathcal{B}_n, P) \in \mathcal{M}_{\vec{p}}$. $\qquad\qquad\square$

Proposition 6.1.12. 1) *If* $\liminf_{k \to \infty} p_k > 1$, *then every martingale* $(f_n, \mathcal{B}_n, P)$ *in* $\mathcal{M}_{\vec{p}}$ *is uniformly integrable.*

2) *If* $\liminf_{k \to \infty} p_k = 1$, *then in* $\mathcal{M}_{\vec{p}}$ *there exist nonuniformly integrable martingales.*

Proof. 1) It is sufficient to carry out the arguments for non-negative $(f_n, \mathcal{B}_n, P)$. The condition of the proposition implies the existence of an index m and a number $\varepsilon > 0$ such that $p_k \geq 1 + \varepsilon$ for $k > m$. It is not difficult to see that P^m-a.s. with respect to $\omega^m \in \Omega^m$, random sequence $(g_n^{\omega^m}, \mathcal{F}_n, P_m)_{n=m+1}^{\infty}$, where

$$g_n^{\omega^m} := f_n(\omega^m, \cdot), \quad \mathcal{F}_n = \{A_m^n \times \Omega_n^{\infty} : A_m^n \in \mathcal{A}_m^n\}$$

is a martingale. We have:

$$\| \sup_{n>m} \|g_n^{\omega^m}\|_{\vec{p}_m} \|_{\vec{p}^m, \omega^m} = \| \lim_{n \to \infty} \uparrow \|f_n(\omega^m, \omega_m)\|_{\vec{p}_m, \omega_m} \|_{\vec{p}^m, \omega^m} =$$

$$= \lim_{n \to \infty} \| \|f_n(\omega^m, \omega_m)\|_{\vec{p}_m, \omega_m} \|_{\vec{p}^m, \omega^m} = \sup_n \|f_n\|_{\vec{p}} < \infty.$$

From this it follows that P^m-a.s. with respect to $\omega^m \in \Omega^m$, $\sup_{n>m} \|g_n^{\omega^m}\|_{\vec{p}_m}$ is finite, and since all coordinates of vector $\vec{p}_m$ are greater than $1 + \varepsilon$, then the martingale

$(g_n^{\omega^m}, \mathscr{F}_n, P_m)$ is uniformly integrable. Therefore, denoting $f := \lim_{n \to \infty} f_n$, we obtain P^m-a.s. with respect to $\omega^m \in \Omega^m$:

$$\int_{\Omega_m} f_n(\omega^m, \omega_m)\, dP_m(\omega_m) = \int_{\Omega_m} f(\omega^m, \omega_m)\, dP_m(\omega_m).$$

Integrating this equation over Ω^m we have: $E[f_n] = E[f]$. And from this it follows (see [96: p. 205, Theorem 5]) that $(f_n, \mathscr{B}_n, P)$ is a uniformly integrable martingale.

2) We shall assume for simplicity that $(\Omega_k, \mathscr{A}_k, Q_k)$ are continuous probability spaces. Without loss of generality we may assume that $\sum_{k=1}^\infty (1 - 1/p_k) < \infty$. Let $A_k \in \mathscr{A}_k$ be such that $Q_k(A_k) = 1/2$. We set

$$B_n := \left(\prod_{k=1}^{n} A_k \right) \times \Omega_n, \quad f_n := 2^n 1_{B_n}.$$

It is obvious that $(f_n, \mathscr{B}_n, P)$ is a non-negative martingale. We have:

$$\sup_n \|f_n\|_{\vec{p}} = \prod_{k=1}^{\infty} 2^{1-1/p_k} = \exp\left(\ln 2 \sum_{k=1}^{\infty} (1 - 1/p_k)\right) < \infty.$$

On the other hand, $P(B_n) \to 0$ for $n \to \infty$, i.e. $f_n \to 0$ a.s. Therefore the martingale $(f_n, \mathscr{B}_n, P)$ is nonuniformly integrable. $\qquad\square$

6.2 Classes $h_{\vec{p}}$ of harmonic functions in the semispace $\mathbb{T}_+^\infty$

6.2.1

In this section we shall consider some properties of the space $\mathcal{M}_{\vec{p}}$ on $\Omega = \mathbb{T}^\infty$ associated with its group structure.

We introduce notation from Sec. 6.1 which will be used here: $\Omega_k = \mathbb{T}_k = \mathbb{T}$, $\mathscr{A}_k$ the σ-algebra of Borel subsets of $\mathbb{T}_k$, $Q_k = dx_k$ the Haar measure on $(\mathbb{T}_k, \mathscr{A}_k)$. Further, $\Omega^n = \mathbb{T}^n$, $\Omega = \mathbb{T}^\infty$, $\Omega_n^\infty = \mathbb{T}_n^\infty = \prod_{k=n+1}^\infty \mathbb{T}_k$, $P = dx$ is the Haar measure on $\mathbb{T}^\infty$, $\nu_n := \varepsilon^n \otimes P_n$, $\nu^n := P^n \otimes \varepsilon_n$ (ε^n and ε_n are the Dirac measures on the neutral elements of groups $\mathbb{T}^n$ and $\mathbb{T}_n^\infty$ respectively), $x := \omega$, $x_k := \omega_k$. We note further that:

$$f \cdot P = f * \nu, \quad f \cdot P^n = f * \nu^n, \quad f \cdot P_n = f * \nu_n.$$

We shall denote the set of cylindrical infinitely differentiable functions on $\mathbb{T}^\infty$ by $\mathscr{D}$. In addition to the properties C.1)–C.11) considered in Sec. 6.1.2, we introduce the following three properties ($1 \le \vec{p} \le \infty$):

C.12) the space $L_{\vec{p}}$ is separable;

C.13) the set $\mathscr{D}$ is dense in $L_{\vec{p}}$;

C.14) any function $f \in L_{\vec{p}}$ is continuous on the whole, i.e. for any $\varepsilon > 0$ there exists a neighborhood V of the neutral element of group $\mathbb{T}^{\infty}$ such that for any $y \in V$ we have

$$\|f(x+y) - f(x)\|_{\vec{p},x} < \varepsilon. \tag{6.2.1}$$

Proposition 6.2.1. *Let* $1 \leq \vec{p} \leq \infty$. *Then*

$$(\mathrm{I}) \Leftrightarrow \mathrm{C}.12) \Leftrightarrow \mathrm{C}.13) \Leftrightarrow \mathrm{C}.14).$$

Proof. (I) $\Leftrightarrow$ C.12). This follows from the general theory of ideal spaces (see [71: p. 142, Theorem 3]), since the Haar measure is obviously separable.

(I) $\Rightarrow$ C.13). Since (I) is satisfied, then from Propositions 6.1.8 and 6.1.9 it follows that $\sup_k p_k < \infty$, and the fulfillment of C.13) becomes obvious.

C.13) $\Rightarrow$ (I). C.13) implies (III). Again from Proposition 6.1.9 it follows that $\sup_k p_k < \infty$, and thus from Proposition 6.1.8 that (I) is satisfied.

C.13) $\Rightarrow$ C.14). Let $g \in \mathscr{D}$ and $\|f - g\|_{\vec{p}} < \varepsilon/3$. then

$$\|f(x+y) - f(x)\|_{\vec{p},x} \leq \|f(x+y) - g(x+y)\|_{\vec{p},x} +$$

$$+\|g(x+y) - g(x)\|_{\vec{p},x} + \|g - f\|_{\vec{p}} \leq \frac{2\varepsilon}{3} + \|g(x+y) - g(x)\|_{\vec{p},x}.$$

Applying the property of uniform continuity to g, we get that there exists a neighborhood V of the neutral element of group $\mathbb{T}^{\infty}$ such that $\|g(x+y) - g(x)\|_{\vec{p},x} < \frac{\varepsilon}{3}$, if $y \in V$.

C.14) $\Rightarrow$ C.13). Without loss of generality, we may assume that the neighborhood V is symmetric. Let a function $\varphi \in \mathscr{D}$ be such that $\varphi \geq 0$, $\operatorname{supp} \varphi \subset V$, and $\int_{\mathbb{T}^{\infty}} \varphi \, dx = 1$. Then $f * \varphi \in \mathscr{D}$ and we have

$$f * \varphi(x) - f(x) = \int_{\mathbb{T}^{\infty}} f(x-y)\varphi(y) \, dy - \int_{\mathbb{T}^{\infty}} f(x)\varphi(y) \, dy =$$

$$= \int_{\mathbb{T}^{\infty}} [f(x-y) - f(x)]\varphi(y) \, dy.$$

Applying the Minkowski integral inequality (Proposition 6.1.5), we get

$$\|f * \varphi - f\|_{\vec{p}} \leq \int_{\mathbb{T}^{\infty}} \|f(x-y) - f(x)\|_{\vec{p},x}\varphi(y) \, dy =$$

$$= \int_V \|f(x-y) - f(x)\|_{\vec{p},x}\varphi(y) \, dy < \varepsilon. \qquad \square$$

Corollary 6.2.1. *Properties C.1)–C.14) are satisfied if and only if* $\sup_k p_k < \infty$. *In this case, for any* $f \in L_{\vec{p}}$ *and for any* δ-*shaped sequence* (φ_n) *we have* $\|f * \varphi_n - f\|_{\vec{p}} \to 0$, $n \to \infty$.

In the case which we are considering, all $(\Omega_k, \mathscr{A}_k)$ are standard Borel spaces, therefore according to [85: Chapter V], space $M_{\vec{p}}$ is naturally identified with the space of signed Borel measures on $\mathbb{T}^\infty$ that are $(\mathscr{B}_n)$-locally absolutely continuous with respect to Haar measure (see [96: p. 511]) and such that the martingales generated by them are in $\mathscr{M}_{\vec{p}}$.

Proposition 6.2.2. (*Young's inequality*) *Let* $1 \le \vec{p}, \vec{q}, \vec{r} \le \infty$, $1/\vec{p} + 1/\vec{q} = 1 + 1/\vec{r}$. *If* $\mu_1 \in \mathscr{M}_{\vec{p}}$, $\mu_2 \in \mathscr{M}_{\vec{q}}$, *then* $\mu_1 * \mu_2 \in \mathscr{M}_{\vec{r}}$, *where*

$$\|\mu_1 * \mu_2\|_{\vec{r}} \le \|\mu_1\|_{\vec{p}} \|\mu_2\|_{\vec{q}}. \tag{6.2.2}$$

Proof. Applying the associativity and commutativity of the convolution operation and the idempotency of measure ν_n, we obtain

$$(\mu_1 * \mu_2) * \nu_n = (\mu_1 * \nu_n) * (\mu_2 * \nu_n).$$

By the definition of $\mathscr{M}_{\vec{p}}$, measures $\mu_1 * \nu_n$ and $\mu_2 * \nu_n$ are absolutely continuous with respect to Haar measure, and their densities are functions of the first n variables. Applying the well-known Young's inequality for such functions (see [28: § 10, Theorem 1]), we get

$$\|(\mu_1 * \mu_2) * \nu_n\|_{\vec{r}^n} \le \|\mu_1 * \nu_n\|_{\vec{p}^n} \|\mu_2 * \nu_n\|_{\vec{q}^n}.$$

Letting $n \to \infty$, we get (6.2.2). $\qquad\square$

Theorem 6.2.1. *Let* $\mathscr{K}$ *be a subset of functions of* $L_{\vec{p}}$ $(1 \le \vec{p} \le \infty)$ *and let the following conditions be satisfied:*
a) $\sup_{f \in \mathscr{K}} \|f\|_{\vec{p}} < \infty$;
b) *for every* $\varepsilon > 0$ *there exists a neighborhood* V *of the neutral element on* $\mathbb{T}^\infty$
   such that for any $f \in \mathscr{K}$, $y \in V$, *inequality* (6.2.1) *is satisfied.*

Then the set $\mathscr{K}$ *is relatively compact in* $L_{\vec{p}}$.
   If $\sup_k p_k < \infty$, *then the relative compactness of* $\mathscr{K}$ *implies the fulfillment of properties* a) *and* b).

Proof. 1) Let conditions a) and b) be satisfied and let $\varphi \in \mathscr{D}$ depend on the first n coordinates, $\varphi \ge 0$, $\operatorname{supp}\varphi \subset V$, and $\int_{\mathbb{T}^\infty} \varphi \, dx = 1$. We shall consider the family of functions:

$$\widetilde{\mathscr{K}} = \{f * \varphi : f \in \mathscr{K}\} \subset \mathbb{C}(\mathbb{T}^n).$$

Applying the Hölder inequality, we get that, first,

$$\sup_{f \in \mathscr{K}} \|f * \varphi\|_{\mathbb{C}(\mathbb{T}^n)} \le \left(\sup_{f \in \mathscr{K}} \|f\|_{\vec{p}} \right) \|\varphi\|_{\vec{q}} < \infty,$$

and second,

$$\|f * \varphi(x + y) - f * \varphi(x)\|_{\infty,x} \le (\sup_{f \in \mathscr{K}} \|f\|_{\vec{p}}) \|\varphi(x + y) - \varphi(x)\|_{\vec{q},x}$$

where in the last inequality the expression on the right is small when y lies in a sufficiently small neighborhood of the neutral element. By Arzelà's theorem we get that $\widetilde{\mathcal{K}}$ is relatively compact in $C(\mathbb{T}^n)$, i.e. $\widetilde{\mathcal{K}}$ has a finite ε-net $\{f_i\}_1^m$ in $C(\mathbb{T}^n)$. We shall show that $\{f_i\}_1^m$ is a 2ε-net of $\mathcal{K}$ in $L_{\vec{p}}$. In fact, for $f \in \mathcal{K}$ we find f_i such that $\|f * \varphi - f_i\|_\infty < \varepsilon$ and we get

$$\|f - f_i\|_{\vec{p}} \leq \|f - f * \varphi\|_{\vec{p}} + \|f * \varphi - f_i\|_{\vec{p}} < 2\varepsilon,$$

since $\|f - f * \varphi\|_{\vec{p}} < \varepsilon$ by inequality (6.2.1) and

$$\|f * \varphi - f_i\|_{\vec{p}} \leq \|f * \varphi - f_i\|_\infty < \varepsilon.$$

2) Let $\sup_k p_k < \infty$ and $\mathcal{K}$ be relatively compact. The fulfillment of property a) is obvious. We shall prove b). Let $\{f_i\}_1^m$ be a finite ε-net of $\mathcal{K}$. Corollary 6.2.1 implies the existence of a neighborhood V of the neutral element such that for any $y \in V$, $i = 1, 2, \ldots$

$$\|f_i(x + y) - f_i(x)\|_{\vec{p},x} < \varepsilon.$$

Finding a function f_i for function $f \in \mathcal{K}$ such that $\|f - f_i\|_{\vec{p}} < \varepsilon$, for any $y \in V$ we get:

$$\|f(x + y) - f(x)\|_{\vec{p},x} \leq \|f(x + y) - f_i(x + y)\|_{\vec{p},x} +$$
$$+ \|f_i(x + y) - f_i(x)\|_{\vec{p},x} + \|f_i - f\|_{\vec{p}} < 3\varepsilon.$$

6.2.2

We shall consider the group $E = \mathbb{R}^1 \times \mathbb{T}^\infty$. We shall denote elements of this group by (t, x), $t \in \mathbb{R}^1$, $x \in \mathbb{T}^\infty$. Let $\mathcal{L} = \sum_{k=0}^\infty a_k \partial_k^2$ $(a_0 = 1, \partial_0 = \partial_t)$ be an elliptic differential operator on E.

By analogy with the finite-dimensional case $E = \mathbb{R}^n$ (see [102]), we introduce into consideration the space $h_{\vec{p}}$ of harmonic functions (= weak solutions of equation $\mathcal{L}u = 0$) in semispace $\mathbb{T}_+^\infty =]0, \infty[\times \mathbb{T}^\infty$. That is, we shall say that a function $u \in L_{1\,\mathrm{loc}}(\mathbb{T}_+^\infty)$ is in the space $h_{\vec{p}}$ if the following conditions are satisfied:

1) the function u is a weak solution of equation $\mathcal{L}u = 0$ on set $\mathbb{T}_+^\infty$;
2) $\|u\|_{h_{\vec{p}}} := \operatorname*{ess\,sup}_{t>0} \|u(t, x)\|_{\vec{p},x} < \infty.$

It is obvious that $h_{\vec{p}}$ is a Banach space. Later it will be shown that under fulfillment of certain conditions the spaces $h_{\vec{p}}$ and $\mathcal{M}_{\vec{p}}$ are isometric. In this connection the mentioned isometry on the one hand is realized by an operator of the solution of the Dirichlet problem, and on the other hand, by an operator associating each function $u \in h_{\vec{p}}$ with its boundary values (understood in the extended sense).

Below we shall denote by (Q_t) the Cauchy semigroup associated with elliptic operator $\mathcal{L}' = \sum_{k=1}^{\infty} a_k \partial_k^2$ on group $\mathbb{T}^{\infty}$ (see Sec. 5.1.1):

$$\widehat{Q_t\varphi}(\theta) = \exp\left(-t\sqrt{\Psi(\theta)}\right),$$

$$\Psi(\theta) = \sum_{k=1}^{\infty} a_k \theta_k^2, \ \theta = (\theta_k)_1^{\infty} \in \mathbb{Z}^{(\infty)}.$$

Proposition 6.2.3. *Let $f \in L_{\vec{p}}$, $1 \le \vec{p} \le \infty$. Set $u_f(t, x) := Q_t f(x)$, then:*
1) $u_f \in h_{\vec{p}}$;
2) $\|u_f\|_{h_{\vec{p}}} = \|f\|_{\vec{p}}$;
3) *if* $\sup_k p_k < \infty$, *then* $\|u_f(t, x) - f(x)\|_{\vec{p},x} \to 0$ *for* $t \to 0$;
4) *if* $\inf_k p_k > 1$, *then* $u_f(t, x) \to f(x)$ *for* $t \to 0$ *(dx-a.e.).*

Proof. 1) Function u_f can be represented as a weak limit of functions u_{f_n}, $f_n \in \mathcal{D}$. It is easy to see that for any n the function u_{f_n} is harmonic on set $\mathbb{T}_{+}^{\infty}$. After passage to the limit we get the harmonicity of function u_f on this set.

The inequality $\sup_{t>0} \|u_f(t, x)\|_{\vec{p},x} \le \|f\|_{\vec{p}}$ follows from Young's inequality (Proposition 6.2.2). We shall prove the converse inequality. According to the well-known finite-dimensional result (see [102]),

$$u_f * \nu_n = u_{f*\nu_n} \to f * \nu_n, \ t \downarrow 0 \ (dx - a.e.)$$

Applying the Fatou property in $L_{\vec{p}}$, we have:

$$\|f * \nu_n\|_{\vec{p}} \le \sup_{t>0} \|u_f * \nu_n(t, x)\|_{\vec{p},x} \le \sup_{t>0} \|u_f(t, x)\|_{\vec{p},x}.$$

From this we get:

$$\|f\|_{\vec{p}} \le \sup_{t>0} \|u_f(t, x)\|_{\vec{p},x}.$$

Thus properties 1) and 2) are proved.

3) The given property follows in a standard manner from the density of $\mathcal{D}$ in $L_{\vec{p}}$ (which occurs since $\sup_k p_k < \infty$).

4) We choose a number r such that $1 < r < \inf_k p_k$. It is obvious that $f \in L_r$. We define the following functions on L_r:

$$Kg(x) = \limsup_{t\to 0} u_g(t, x) - \liminf_{t\to 0} u_g(t, x)$$

$$Mg(x) = \sup_{t>0} |u_g(t, x)|.$$

With respect to function Mg, the following inequality is known (see [79: p. 167]):

$$\|Mg\|_r \le c_r \|g\|_r. \tag{6.2.3}$$

Since $Kg \le 2Mg$, then using simple calculations and inequality (6.2.3), for any $\varepsilon > 0$ we get:

$$\int_{\mathbb{T}^\infty} 1_{\{Kg>\varepsilon\}} dx \le \int_{\mathbb{T}^\infty} 1_{\{2Mg>\varepsilon\}} dx \le \left(\frac{2}{\varepsilon}\right)^r \|Mg\|_r^r \le \left(\frac{2}{\varepsilon}\right)^r c_r \|g\|_r^r.$$

Since $\mathcal{D}$ is dense in L_r, it is possible to write $f = h + g$, where $h \in \mathcal{D}$ and $\|g\|_r < \delta$. It is obvious that $Kh = 0$. Therefore $Kf \le Kh + Kg = Kg$, and therefore

$$\int_{\mathbb{T}^\infty} 1_{\{Kf>\varepsilon\}} dx \le c(\varepsilon, r)\|g\|_r^r \le c(\varepsilon, r)\delta^r.$$

Since δ can be chosen as small as necessary, we get $Kf = 0$, (dx-a.e.), i.e. $\lim_{t\to 0} u_f(t, x)$ exists (dx-a.e.). And finally, from part 3) of this proposition it follows that this limit coincides with f (dx-a.e.). $\qquad\square$

Proposition 6.2.4. *Let a function u defined on the set $\mathbb{T}_+^\infty$ be in $h_{\vec{p}}$. Then on $\mathbb{T}^\infty$ there exists a measure μ such that for all $t > 0$ the measures $Q_t\mu$ are absolutely continuous with respect to Haar measure and $u(t, x) = Q_t\mu(x)$ ($dt \otimes dx$-a.e.).*
If in this connection $p_k > 1$, $\forall k \ge 1$, then $\mu \in \mathcal{M}_{\vec{p}}$.

Proof. We turn to the proof of the second part of the proposition, since the same method is used in the proof of the first part. We denote $u_t(x) := u(t, x)$ and consider the function $u_t^n = u_t * v_n$, which depends on $n + 1$ variables. The function $u_t^n(x)$ is harmonic on set $\mathbb{T}_+^n =]0, \infty[\times \mathbb{T}^n$ and by Weyl's lemma has continuous regularization $\tilde{u}_t^n(x)$. We have

$$\sup_{t>0} \|\tilde{u}_t^n(x)\|_{\vec{p}} = \operatorname{ess\,sup}_{t>0} \|u_t^n(x)\|_{\vec{p}} \le \operatorname{ess\,sup}_{t>0} \|u_t\|_{\vec{p}} := c.$$

Considering that $\inf_{1\le k\le n} p_k > 1$, and applying the known finite-dimensional result (see [102]), we get the existence of a function f_n on $\mathbb{T}^n$ such that for all $t > 0$ we have $\tilde{u}_t^n = Q_t f_n$ (dx-a.e.), where

$$\|f_n\|_{\vec{p}} = \sup_{t>0} \|Q_t f_n\|_{\vec{p}} = \sup_{t>0} \|\tilde{u}_t^n\|_{\vec{p}} \le c.$$

Thus we have obtained a martingale $(f_n, \mathcal{B}_n, dx)$ which is in $\mathcal{M}_{\vec{p}}$. Therefore there exists a measure μ on $\mathbb{T}^\infty$ which is $(\mathcal{B}_n)$-locally absolutely continuous, where $\mu|_{\mathcal{B}_n} = f_n dx$. Now without difficulty we shall verify that, first, measure $Q_t\mu$ is absolutely continuous with respect to Haar measure, and second, that $Q_t\mu(x) = u(t, x)$ ($dt \otimes dx$-a.e.). $\qquad\square$

Corollary 6.2.2. *In each of the following two cases the spaces $h_{\vec{p}}$ and $\mathcal{M}_{\vec{p}}$ are isometric:*

1) $\inf_k p_k > 1$;

2) $p_k > 1$, $k = 1, 2, \ldots$, *and for all $t > 0$ the series $\sum_{k=1}^\infty e^{-a_k t}$ converges.*

In the general case the space $h_{\vec{p}}$ can be isometrically imbedded in $\mathcal{M}_{\vec{p}}$.

Proof. 1) In this case $\mathcal{M}_{\vec{p}} = L_{\vec{p}}$ and it remains to use Propositions 6.2.3 and 6.2.4. 2) From convergence for all $t > 0$ of the series $\sum_{k=1}^\infty e^{-a_k t}$ it follows that for all $t > 0$ we have $Q_t(x, dy) = q_t(x - y)dy$ (see Proposition 4.2.1 and Theorem 4.3.2). Therefore for any measure $\mu \in \mathcal{M}_{\vec{p}}$ the measure $Q_t \mu$ is absolutely continuous with respect to Haar measure, and its density $Q_t \mu(x)$ is in $L_{\vec{p}}$. Without difficulty it can be verified that the function $u_\mu(t, x) = Q_t \mu(x)$ is in $h_{\vec{p}}$ and $\|u_\mu\|_{h_{\vec{p}}} = \|\mu\|_{\vec{p}}$. Now it remains to use Proposition 6.2.4. $\qquad\square$

6.3 $\mathcal{M}_{\vec{p}}$-estimates of potentials. Sobolev inequality on the group $\mathbb{T}^\infty$

In the following, we shall think of the one-dimensional torus $\mathbb{T}$ as the segment $[-\pi, \pi]$ with the ends identified (in this connection the Haar measure is simply the normed Lebesgue measure). Let $n_t(x)$ be the density of the normal distribution on $\mathbb{T}$ with parameters $(0, 2t)$:

$$n_t(x) = \sqrt{\frac{\pi}{t}} \sum_{k \in \mathbb{Z}} \exp\left\{ -\frac{(x - 2\pi k)^2}{4t} \right\}.$$

We introduce into consideration the measures $n_{a_k t}(dx) = n_{a_k t}(x)dx$, $a_k > 0$, and we form the Gaussian semigroup (μ_t) on $\mathbb{T}^\infty$:

$$\mu_t = \otimes_{k=1}^\infty n_{a_k t}.$$

For every $t > 0$ the measure μ_t is $\mathcal{B}_n$-locally absolutely continuous with respect to Haar measure on $\mathbb{T}^\infty$. This implies that $\mu_t \in \mathcal{M}_1$. For additional conditions on coefficients a_k and vector $\vec{p}$, a stronger result has been verified:

Proposition 6.3.1. *Let $1 \le \vec{p} \le \infty$, $1/\vec{p} + 1/\vec{q} = 1$. We set $\tilde{a}_k := \min\{a_k, 1/\log(k+1)\}$, $k = 1, 2, \ldots$, and let the following condition be satisfied:*

$$\prod_{k=1}^\infty \tilde{a}_k^{1/q_k} > 0. \tag{6.3.1}$$

Then:

1) $\mu_t \in \mathcal{M}_{\vec{p}}$ $(\forall t > 0)$;
2) *the series* $\alpha := \sum_{k=1}^{\infty} 1/q_k$ *converges and for any* $\varepsilon > 0$ *the following assymptotic inequalities are satisfied:*

$$t^{-(\alpha/2-\varepsilon)} \prec ||\mu_t||_{\vec{p}} \le t^{-\alpha/2}, \ t \downarrow 0.$$

Remark 6.3.1. Let condition (6.3.1) be satisfied. Then from the obvious inequalities

$$1 > \prod_{k=2}^{\infty} (1/\ln k)^{1/q_k} \ge \prod_{k=2}^{\infty} \tilde{a}_k^{1/q_k} > 0$$

we deduce that the infinite product $\prod_{k=2}^{\infty} (\ln k)^{1/q_k}$ converges. The obtained property, in turn, is equivalent to the condition

$$\sum_{k=2}^{\infty} \ln \ln k / q_k < \infty. \tag{6.3.2}$$

In particular, we get the convergence of series $\alpha = \sum_{k=1}^{\infty} 1/q_k$ from condition (6.3.1).

And finally, we note one particular case: let the sequence $\{a_k\}$ satisfy the condition

$$\liminf_{k \to \infty} a_k \ln k = c > 0, \tag{6.3.3}$$

then condition (6.3.1) reduces to condition (6.3.2). $\qquad\qquad\square$

The proof of the proposition is essentially based on some estimates of function $t \to ||n_t||_p$ which are given below.

Lemma 6.3.1. *The following properties hold true:*

1) $||n_t||_p \le \left(\sqrt{\frac{\pi}{t}}\right)^{\frac{1}{q}} \left(\frac{1}{p}\right)^{\frac{1}{2p}} \left[1 + e^{-\frac{\pi^2}{8t}}\right], \ 0 < t < 1/2;$

2) $||n_t||_p \sim \left(\sqrt{\frac{\pi}{t}}\right)^{\frac{1}{q}} \left(\frac{1}{p}\right)^{\frac{1}{2p}}, \ t \downarrow 0;$

3) $||n_t||_p - 1 \sim \frac{p}{2q} e^{-2t}, \ t \uparrow \infty.$

Proof. We have:

$$||n_t||_p \le \sqrt{\frac{\pi}{t}} \sum_{k \in \mathbb{Z}} \left\|e^{-\frac{(x-2\pi k)^2}{4t}}\right\|_{p,x}.$$

We estimate each term on the right-hand side of the inequality:

$$\left\|e^{-\frac{x^2}{4t}}\right\|_{p,x}^{p} = \frac{1}{2\pi} \int_{-\pi}^{\pi} e^{-\frac{x^2 p}{4t}} dx \le \frac{1}{2\pi} \int_{-\infty}^{\infty} e^{-\frac{x^2 p}{4t}} dx = \sqrt{\frac{t}{\pi}} \left(\frac{1}{p}\right)^{\frac{1}{2}}$$

$$\left\| e^{-\frac{(x-2\pi k)^2}{4t}} \right\|_{p,x}^p = \frac{1}{2\pi} \int_{-\pi}^{\pi} e^{-\frac{(x-2\pi k)^2 p}{4t}} \, dx = \frac{1}{2\pi} \int_{-\pi(2k+1)}^{-\pi(2k-1)} e^{-\frac{x^2 p}{4t}} \, dx \le$$

$$\le \begin{cases} e^{-\frac{\pi^2(2k-1)^2 p}{4t}}, & k \ge 1 \\ e^{-\frac{\pi^2(2k+1)^2 p}{4t}}, & k \le -1. \end{cases}$$

From this we get for $0 < t < 1/2$:

$$\|n_t\|_p \le \sqrt{\frac{\pi}{t}} \left(\left(\sqrt{\frac{t}{\pi}} \right)^{\frac{1}{p}} \left(\frac{1}{p} \right)^{\frac{1}{2p}} + 2 \sum_{k=0}^{\infty} e^{-\frac{\pi^2(2k+1)^2}{4t}} \right) \le$$

$$\le \sqrt{\frac{\pi}{t}} \left(\left(\sqrt{\frac{t}{\pi}} \right)^{\frac{1}{p}} \left(\frac{1}{p} \right)^{\frac{1}{2p}} + 2 \sum_{k=1}^{\infty} e^{-\frac{\pi^2 k}{4t}} \right) =$$

$$= \sqrt{\frac{\pi}{t}} \left(\left(\sqrt{\frac{t}{\pi}} \right)^{\frac{1}{p}} \left(\frac{1}{p} \right)^{\frac{1}{2p}} + 2 e^{-\frac{\pi^2}{4t}} \left(1 - e^{-\frac{\pi^2}{4t}} \right)^{-1} \right) \le$$

$$\le \left(\sqrt{\frac{\pi}{t}} \right)^{\frac{1}{q}} \left(\frac{1}{p} \right)^{\frac{1}{2p}} \left(1 + 3 p^{\frac{1}{2p}} \left(\sqrt{\frac{\pi}{t}} \right)^{\frac{1}{p}} e^{-\frac{\pi^2}{4t}} \right) \le$$

$$\le \left(\sqrt{\frac{\pi}{t}} \right)^{\frac{1}{q}} \left(\frac{1}{p} \right)^{\frac{1}{2p}} \left(1 + e^{-\frac{\pi^2}{8t}} \right).$$

2) We apply inequality $(a + b)^p \ge a^p + b^p$ $(a, b \ge 0, \ p \ge 1)$:

$$\|n_t\|_p \ge \left(\sqrt{\frac{\pi}{t}} \right) \left(\sum_{k \in \mathbb{Z}} \left\| e^{-\frac{(x-2\pi k)^2}{4t}} \right\|_{p,x}^p \right)^{1/p} =$$

$$= \left(\sqrt{\frac{\pi}{t}} \right) \left(\sum_{k \in \mathbb{Z}} \frac{1}{2\pi} \int_{-\pi}^{\pi} e^{-\frac{(x-2\pi k)^2 p}{4t}} \, dx \right)^{1/p} =$$

$$= \left(\sqrt{\frac{\pi}{t}} \right) \left(\frac{1}{2\pi} \int_{-\infty}^{\infty} e^{-\frac{x^2 p}{4t}} \, dx \right)^{1/p} = \left(\sqrt{\frac{\pi}{t}} \right)^{\frac{1}{q}} \left(\frac{1}{p} \right)^{\frac{1}{2p}}$$

Thus together with 1) we obtain for $0 < t < 1/2$:

$$\left(\sqrt{\frac{\pi}{t}} \right)^{\frac{1}{q}} \left(\frac{1}{p} \right)^{\frac{1}{2p}} \le \|n_t\|_p \le \left(\sqrt{\frac{\pi}{t}} \right)^{\frac{1}{q}} \left(\frac{1}{p} \right)^{\frac{1}{2p}} \left(1 + e^{-\frac{\pi^2}{8t}} \right)$$

which also implies the truth of property 2).

3) For $t > 0$ and $x \in [-\pi, \pi]$ we denote

$$\varphi(t, x) := \sum_{k=1}^{\infty} e^{-tk^2} \cos kx.$$

Then we shall have:

1) $n_t(x) = \sum_{k \in \mathbb{Z}} e^{-tk^2} \cos kx = 1 + 2\varphi(t, x)$,

2) $|\varphi(t, x)| \le \sum_{k=1}^{\infty} e^{-tk^2} \le \int_0^{\infty} e^{-tx^2} dx = \frac{1}{2}\sqrt{\frac{\pi}{t}}$.

For $t > \pi$ we get from (2):

$$|\varphi(t, x)| \le \frac{1}{2}\sqrt{\frac{\pi}{t}} < \frac{1}{2},$$

therefore the series written below converges uniformly for $x \in [-\pi, \pi]$:

$$(n_t(x))^p = \sum_{k=0}^{\infty} \binom{p}{k} 2^k \varphi(t, x)^k.$$

After termwise integration we get

$$\|n_t\|_p^p = 1 + \sum_{k=2}^{\infty} \binom{p}{k} 2^k \frac{1}{2\pi} \int_{-\pi}^{\pi} \varphi(t, x)^k dx.$$

Further, we have:

$$e^t \varphi(t, x) = \cos x + \sum_{k=2}^{\infty} e^{-tk(k-1)} \cos kx = \cos x + A(t, x),$$

$$|A(t, x)| \le \sum_{k=1}^{\infty} e^{-tk^2} \le \frac{1}{2}\sqrt{\frac{\pi}{t}},$$

therefore uniformly with respect to $x \in [-\pi, \pi]$

$$\lim_{t \to \infty} e^t \varphi(t, x) = \cos x.$$

Therefore

$$\lim_{t \to \infty} e^{2t}(\|n_t\|_p^p - 1) = \frac{p(p-1)}{2},$$

and consequently

$$\|n_t\|_p = \left[1 + \frac{p(p-1)}{2} e^{-2t}(1 + o(1))\right]^{1/p}.$$

Finally we get for $t \uparrow \infty$

$$\|n_t\|_p - 1 \sim \frac{1}{p} \cdot \frac{p(p-1)}{2} e^{-2t} = \frac{p}{2q} e^{-2t}.$$

The proof of the lemma is complete.

Proof of Proposition 6.3.1. From Young's inequality it follows that the function $t \to \|n_t\|_p$ decreases monotonically, and therefore

$$\|n_{a_k t}\|_p \le \|n_{\widetilde{a_k t}}\|_p.$$

Using inequality 1) from Lemma 6.3.1, we have from the definition of measure μ_t and its mixed norm $(0 < t < 1/2)$:

$$\|\mu_t\|_{\vec{p}} = \prod_{k=1}^{\infty} \|n_{a_k t}\|_{p_k} \leq \prod_{k=1}^{\infty} \|n_{\widetilde{a}_k t}\|_{p_k} \leq$$

$$\leq \prod_{k=1}^{\infty} \left(\sqrt{\frac{\pi}{\widetilde{a}_k t}}\right)^{\frac{1}{q_k}} \prod_{k=1}^{\infty} \left(\frac{1}{p_k}\right)^{\frac{1}{2p_k}} \prod_{k=1}^{\infty} \left(1 + e^{-\pi^2/8\widetilde{a}_k t}\right) =$$

$$= \prod_1 \ \prod_2 \ \prod_3$$

From condition (6.3.1) we get

$$\prod_1 = \left(\prod_{k=1}^{\infty} \widetilde{a}_k^{1/q_k}\right)^{-\frac{1}{2}} \left(\frac{\pi}{t}\right)^{\frac{1}{2}\sum_{k=1}^{\infty} 1/q_k} = c_1 t^{-\alpha/2},$$

$$\prod_2 = \left(\prod_{k=1}^{\infty}\left(1 - \frac{1}{q_k}\right)^{\frac{1}{p_k}}\right)^{\frac{1}{2}} < e^{\alpha/2} := c_2,$$

$$\prod_3 \leq \exp\sum_{k=1}^{\infty} e^{-\pi^2/8\widetilde{a}_k t} \leq \exp\sum_{k=1}^{\infty} e^{-\frac{\pi^2}{8t}\ln k} =$$

$$= \exp\sum_{k=1}^{\infty} \frac{1}{k^{\pi^2/8t}} \leq \exp\sum_{k=1}^{\infty} \frac{1}{k^{3/2}} := c_3.$$

Finally we have for $0 < t < 1/2$:

$$\|\mu_t\|_{\vec{p}} \leq c_4 t^{-\alpha/2}.$$

On the other hand, for the given $\varepsilon > 0$ we choose an index $n \geq 1$ such that $\sum_{k \geq n} 1/q_k < \varepsilon$, then according to asymptotics 2) of Lemma 6.3.1, we get:

$$\|\mu_t\|_{\vec{p}} = \prod_{k=1}^{\infty} \|n_{a_k t}\|_{p_k} \geq \prod_{k=1}^{n} \|n_{a_k t}\|_{p_k} \sim$$

$$\sim c_5 t^{-\frac{1}{2}\sum_{k=1}^{n} 1/q_k} \geq c_5 t^{-(\alpha/2 - \varepsilon)}.$$

The proof of the proposition is complete. $\qquad\qquad\square$

Remark 6.3.2. Let condition (6.3.2) be satisfied, along with condition

$$0 < \prod_{k=1}^{\infty} a_k^{1/q_k} < \infty, \qquad\qquad (6.3.4)$$

then for $0 < t < 1/2$ the following inequality holds:

$$c_1 t^{-\alpha/2} \leq \|\mu_t\|_{\vec{p}} \leq c_2 t^{-\alpha/2}.$$

Indeed, from conditions (6.3.2) and (6.3.4) it follows that condition (6.3.1) is satisfied. Thus, by Proposition 6.3.1, the inequality on the right holds. The inequality on the left also holds, since:

$$||\mu_t||_{\vec{p}} = \prod_{k=1}^{\infty} ||n_{a_k t}||_{p_k} \geq \prod_{k=1}^{\infty} \left(\sqrt{\frac{\pi}{a_k t}}\right)^{\frac{1}{q_k}} \left(\frac{1}{p_k}\right)^{\frac{1}{2p_k}} =$$

$$= \left(\prod_{k=1}^{\infty} a_k^{1/q_k}\right)^{-1/2} \pi^{\alpha/2} \left(\prod_{k=1}^{\infty} \frac{1}{p_k}\right)^{1/2} t^{-\alpha/2} = c_1 t^{-\alpha/2}$$

$\square$

Now we pass to $\mathcal{M}_{\vec{p}}$-estimates of Bessel potentials. We recall (see Sec. 5.4) that Bessel potential $\mathcal{J}_\alpha[f]$ of function f is defined by the equality $\mathcal{J}_\alpha[f] = \mathcal{J}_\alpha * f$, where

$$\mathcal{J}_\alpha(dx) = \frac{1}{\Gamma(\alpha/2)} \int_0^{\infty} e^{-t} t^{\alpha/2-1} \mu_t(dx) dt.$$

Proposition 6.3.2. *Let* $1 \leq \vec{p}, \vec{q} \leq \infty$, $1/\vec{p} + 1/\vec{q} = 1$. *We shall assume that together with condition* (6.3.1) *the following condition is satisfied:*

$$\beta := \sum_{k=1}^{\infty} 1/q_k < \alpha, \tag{6.3.5}$$

then $\mathcal{J}_\alpha \in \mathcal{M}_{\vec{p}}$.

Proof. We shall write the semigroup property for (μ_t) in the form $\mu_{t+\Delta} = \mu_t * \mu_\Delta$. Applying Young's inequality, we get

$$||\mu_{t+\Delta}||_{\vec{p}} \leq ||\mu_t||_{\vec{p}} ||\mu_\Delta||_1 = ||\mu_t||_{\vec{p}}.$$

Thus $t \to ||\mu_t||_{\vec{p}}$ is a decreasing function. Now applying the Minkowski inequality for $(\mathcal{B}_n)$-locally absolutely continuous measures (this inequality easily follows from Proposition 6.1.5) and Proposition 6.3.1, we get:

$$||\mathcal{J}_\alpha||_{\vec{p}} \leq \frac{1}{\Gamma(\alpha/2)} \int_0^{\infty} e^{-t} t^{\alpha/2-1} ||\mu_t||_{\vec{p}} dt \leq$$

$$\leq c_1 \int_0^1 e^{-t} t^{(\alpha-\beta)/2-1} dt + c_2 \int_1^{\infty} e^{-t} t^{\alpha/2-1} dt < \infty.$$

$\square$

Remark 6.3.3. 1) From condition (6.3.5) of the proposition it follows that components p_k of vector $\vec{p}$ tend to one for $k \to \infty$. Here we wish to note that the measure $\mathcal{J}_\alpha \notin \mathcal{M}_{\vec{r}}$, if vector $\vec{r}$ is such that $\inf_k r_k := r > 1$. Indeed, if for such $\vec{r}$ the property $\mathcal{J}_\alpha \in \mathcal{M}_{\vec{r}}$ took place, then, first, the measure $\mathcal{J}_\alpha$ would be absolutely continuous with respect to Haar measure (see Proposition 6.1.12) and second, its density $\mathcal{J}_\alpha(x) \in L_r \subset L_\Delta$ for some $1 < \Delta \le \min(r, 2)$. But then by the Hausdorff–Young Theorem, $\widehat{\mathcal{J}}_\alpha \in L_s(\mathbb{Z}^{(\infty)})$ $(1/s + 1/\Delta = 1)$. Since $\widehat{\mathcal{J}}_\alpha = (1 + \Psi)^{-\alpha/2}$, then this would mean that for some number $\beta > 0$, the function $(1 + \Psi)^{-\beta/2}$ was integrable on infinite-dimensional lattice $\mathbb{Z}^{(\infty)}$ - which is impossible. Indeed, let $n \ge \beta$ and $\mathbb{Z}^{(n)} = \{\theta \in \mathbb{Z}^{(\infty)} : \theta_k = 0, k \ge n\}$; then:

$$\int_{\mathbb{Z}^{(\infty)}} (1 + \Psi(\theta))^{-\beta/2} d\theta \ge \int_{\mathbb{Z}^{(n)}} (1 + \Psi(\theta))^{-\beta/2} d\theta$$

$$= \sum_{\theta_1,\ldots,\theta_n=-\infty}^{\infty} \left[1 + a_1\theta_1^2 + \ldots + a_n\theta_n^2\right]^{-\beta/2}$$

$$\ge c_1 \sum_{\theta_1,\ldots,\theta_n=-\infty}^{\infty} \left[1 + \theta_1^2 + \ldots + \theta_n^2\right]^{-n/2}$$

$$= c_1 \sum_{\theta_1,\ldots,\theta_{n-1}=-\infty}^{\infty} \sum_{\theta_n=-\infty}^{\infty} \left[1 + \theta_1^2 + \ldots + \theta_n^2\right]^{-n/2}$$

$$\ge c_1 \sum_{\theta_1,\ldots,\theta_{n-1}=-\infty}^{\infty} \int_1^{\infty} \left[1 + \theta_1^2 + \ldots + \theta_{n-1}^2 + \theta^2\right]^{-n/2} d\theta$$

$$\ge c_2 \sum_{\theta_1,\ldots,\theta_{n-1}=-\infty}^{\infty} \left[1 + \theta_1^2 + \ldots + \theta_{n-1}^2\right]^{-\frac{n-1}{2}}$$

$$\ge \ldots \ge c_n \sum_{\theta_1=-\infty}^{\infty} \left[1 + \theta_1^2\right]^{-\frac{1}{2}} = \infty.$$

2) Absolute continuity of all measures $\{\mathcal{J}_\alpha\}_{\alpha>0}$ with respect to Haar measure is equivalent to absolute continuity of all measures $(\mu_t)_{t>0}$ with respect to Haar measure (the proof is the same as in Proposition 4.2.1). In turn, absolute continuity of all measures $(\mu_t)_{t>0}$ is equivalent to absolute continuity of one measure $r^1 = \mathcal{J}_2$ (value of resolvent $(r^\lambda)_{\lambda>0}$ for $\lambda = 1$). Thus, in correspondence with Theorem 4.3.2, absolute continuity of all measures $\{\mathcal{J}_\alpha\}_{\alpha>0}$ with respect to Haar measure takes place if and only if the series $\sum_{k=1}^{\infty} e^{-a_k t}$ converges for all $t > 0$. In particular, if this series converges, then by the conditions of Proposition 6.3.2, we can assert that $\mathcal{J}_\alpha \in L_{\vec{p}}$. In contrast to this, for $a_1 = a_2 = \ldots = 1$ all measures $\{\mathcal{J}_\alpha\}_{\alpha>0}$ are *singular* with respect to Haar measure (this property follows from Example 1 (Sec. 5.3.2)) and consequently in this case $\mathcal{J}_\alpha \notin L_{\vec{p}}$.

Theorem 6.3.1. (*infinite-dimensional variant of Sobolev inequality*) *Let* $1 \le \vec{p}, \vec{q}, \vec{\Delta} \le \infty$, $1/\vec{p} = 1/\vec{q} - 1/\vec{\Delta}$ *and* $\tilde{a}_k = \min(a_k, 1/\ln(k+1))$, $k = 1, 2, \ldots$. *We shall assume that the following conditions are satisfied:*

1) $\prod_{k=2}^{\infty} \tilde{a}_k^{1/\Delta_k} > 0$;

2) $\sum_{k=1}^{\infty} 1/\Delta_k < \alpha,$

then the following inequality holds:

$$\|\mathcal{J}_\alpha[f]\|_{\vec{p}} \le c\|f\|_{\vec{q}}.$$

Proof. According to Proposition 6.3.2, measure $\mathcal{J}_\alpha$ is in $\mathcal{M}_{\vec{\Delta}'}$, $1/\vec{\Delta}' + 1/\vec{\Delta} = 1$. Now applying Young's inequality to the function $\mathcal{J}_\alpha[f] = \mathcal{J}_\alpha * f$, $f \in L_{\vec{q}}$, we get the required inequality. $\square$

We note that in the theorem $\vec{p} > \vec{q}$, i.e. the Bessel potential $\mathcal{J}_\alpha[f]$ improves the properties of functions from $L_{\vec{q}}$ $(1 \le \vec{q} \le \infty)$.

For a vector $1 \le \vec{p} \le \infty$ and a number $\alpha \ge 0$ we set:

$$L_{\vec{p}}^\alpha = \left\{ \begin{array}{ll} L_{\vec{p}}, & \alpha = 0 \\ \{\mathcal{J}_\alpha[\varphi] : \varphi \in L_{\vec{p}}\}, & \alpha > 0. \end{array} \right.$$

We denote by $\mathcal{D}_{\vec{p}}$ the closure of set $\mathcal{D}$ of cylindrical infinitely differentiable functions in $L_{\vec{p}}$ (see Proposition 6.2.1 and Corollary 6.2.1) and we set $\mathcal{D}_{\vec{p}}^\alpha = \{\mathcal{J}_\alpha[\varphi] : \varphi \in \mathcal{D}_{\vec{p}}\}$.

Theorem 6.3.2. *Let the vector $1 \le \vec{p} \le \infty$, the number $\alpha \ge 0$, and the sequence $\{a_k\}_1^\infty$ satisfy the conditions:*
1) $\prod_{k=1}^\infty \tilde{a}_k^{1/p_k} > 0$, $\tilde{a}_k := \min(a_k, 1/\ln(k+1))$, $k = 1, 2, \ldots$;
2) $\sum_{k=1}^\infty 1/p_k < \alpha$.
Then:
1. If $\sum_{k=1}^\infty e^{-a_k t} < \infty$ $(\forall t > 0)$, then every function $f \in L_{\vec{p}}^\alpha$ coincides $(dx$-a.e.)
with some continuous function $\tilde{f}$.
2. If $\sum_{k=1}^\infty e^{-a_k t} = \infty$ $(\exists t > 0)$, then every function $f \in \mathcal{D}_{\vec{p}}^\alpha$ coincides $(dx$-a.e.)
with some continuous function $\tilde{f}$.

Proof. Let $f = \mathcal{J}_\alpha[\varphi]$. We choose a number $\tilde{\alpha}$ such that $\sum_{k=1}^\infty 1/p_k < \tilde{\alpha} < \alpha$ and we write

$$f = \mathcal{J}_\Delta[\mathcal{J}_{\tilde{\alpha}}[\varphi]], \Delta := \alpha - \tilde{\alpha}.$$

By Theorem 6.3.1 the function $\mathcal{J}_{\tilde{\alpha}}[\varphi] \in L_\infty$; let $\tilde{\varphi}$ be a Borel function such that $\tilde{\varphi} = \mathcal{J}_{\tilde{\alpha}}[\varphi]$ $(dx$-a.e.). We set $\tilde{f} = \mathcal{J}_\Delta[\tilde{\varphi}]$, then $f = \tilde{f}$ $(dx$-a.e.). We shall show that the function $\tilde{f}$ is continuous. In fact, the convergence of series $\sum_{k=1}^\infty e^{-a_k t}$ $(\forall t > 0)$ implies (Theorem 4.3.2) the absolute continuity with respect to Haar measure of all measures $(\mu_t)_{t>0}$, and therefore, in turn, the absolute continuity with respect to Haar measure of all measures $(\mathcal{J}_\alpha)_{\alpha>0}$. In particular, measure $\mathcal{J}_\Delta(dx) = \mathcal{J}_\Delta(x)dx$, $\mathcal{J}_\Delta(x) \in L_1$. Therefore the function $\tilde{f} = \mathcal{J}_\Delta * \tilde{\varphi}$ $(\tilde{\varphi} \in \mathbb{B})$ is continuous.

2. Let $\{\varphi_n\}_1^\infty \subset \mathcal{D} : \varphi_n \to \varphi$ in $L_{\vec{p}}$ for $n \to \infty$. We shall consider the sequence $\{f_n\}_1^\infty$ of continuous functions: $f_n = \mathcal{J}_\alpha[\varphi_n]$. According to Theorem 6.3.1,

$$\|f_n - f_m\|_\infty \le c\|\varphi_n - \varphi_m\|_{\vec{p}} \to 0, \quad m, n \to \infty.$$

Thus the sequence $\{f_n\}_1^\infty$ is a Cauchy sequence in $\mathbb{C}(\mathbb{T}^\infty)$; we denote the limit of this sequence by $\widetilde{f}$. It is easy to see that $\widetilde{f} = f$ (dx-a.e.). $\qquad\square$

Let V be an open subset of T^∞; we set:

$$\mathcal{D}(V) = \{\alpha \in \mathcal{D} : \operatorname{supp}\alpha \subset V\},$$
$$L_{\vec{p}\,\mathrm{loc}}(V) = \{f : \forall \alpha \in \mathcal{D}(V)\ \alpha f \in L_{\vec{p}}\},$$
$$\mathcal{D}_{\vec{p}\,\mathrm{loc}}(V) = \{f : \forall \alpha \in \mathcal{D}(V)\ \alpha f \in \mathcal{D}_{\vec{p}}\}.$$

In addition to Theorems 5.2.1 and 5.2.2, we give the following

Proposition 6.3.3. *Let vector $1 < \vec{p} < \infty$ and sequence $\{a_k\}_1^\infty$ satisfy the following conditions:*

1) $\prod_{k=2}^\infty \widetilde{a}_k^{1/p_k} > 0$, $\widetilde{a}_k = \min(a_k, 1/\ln(k+1))$, $k = 1, 2, \dots$,

2) $\sum_{k=1}^\infty 1/p_k < 2$,

and let u be a weak solution of equation $\mathcal{L}u = 0$ on V (see Sec. 5.2.1). Then:

1. *If $\sum_{k=1}^\infty e^{-a_k t} < \infty$ ($\forall t > 0$), and function u and its weak derivatives $\partial.u$ are in $L_{\vec{p}\,\mathrm{loc}}(V)$, then there exists a continuous function $\widetilde{u}$ such that $\widetilde{u} = u$ (dx-a.e.) on V.*

2. *If $\sum_{k=1}^\infty e^{-a_k t} = \infty$ ($\exists t > 0$), and function u and its weak derivatives $\partial.u$ are in $\mathcal{D}_{\vec{p}\,\mathrm{loc}}(V)$, then there exists a continuous function $\widetilde{u}$ such that $\widetilde{u} = u$ (dx-a.e.) on V.*

Proof. We shall choose a function $\beta \in \mathcal{D}(V)$ and set $u_\beta = \beta \cdot u$. After simple transformations we get:

$$\mathcal{L}u_\beta - u_\beta = g, \text{ where } g = (\mathcal{L}\beta - \beta)u + 2\sum_{k=1}^\infty a_k \partial_k \beta \cdot \partial_k u.$$

Therefore $u_\beta = -R^1 g = -\mathcal{J}_2[g]$. Now it remains to use Theorem 6.3.2. $\qquad\square$

Remark 6.3.4. It would be very interesting to extend the results of Secs. 5.1, 5.2 and 5.3 to the case of the scale of spaces $\{\mathcal{M}_{\vec{p}}\}$. In particular, the following variant of Theorem 5.1.7 seems to be true.

Theorem. For $1 < \vec{p} < \infty$, the following inequality holds:

$$\||\vec{R}f|\|_{\mathcal{M}_{\vec{p}}} \le C_{\vec{p}}\|f\|_{\mathcal{M}_{\vec{p}}}.$$

This theorem would give a powerful tool for the investigation of weak solutions of the equation $\mathcal{L}u - \lambda u = -f$ in the singular case $X \notin \mathcal{B}$ (see Theorem 5.3.1).

Chapter 7
Some thoughts on probability and analysis on locally compact groups

We consider some problems concerning probabilities and potentials on locally compact groups. The three main objects of investigation are Gaussian measures and weakly continuous convolution semigroups generated by them; Laplacians; and harmonic functions on Lie projective groups. The following questions are under consideration: the dichotomy problem for Gaussian measures; the hypoellipticity problem for the Laplacian; existence and smoothness of the transition density for Brownian motion; its asymptotic behavior for small times and the connection of this asymptotic with the behavior of the eigenvalues of the Laplacian; and inverse problems.

7.1 Dichotomy problem

It is well known that every Gaussian measure on the Euclidean space $\mathbb{R}^n$ is supported on some hypersurface of $\mathbb{R}^n$ and has a strictly positive density with respect to Lebesgue measure there. This simple fact leads to the following important property of Gaussian measures on the Euclidean space $\mathbb{R}^n$: two given Gaussian measures μ and ν are either equivalent or singular: $\mu \sim \nu$ or $\mu \perp \nu$. The same property holds for Gaussian measures on the infinite dimensional Hilbert space (Hajek–Feldman Dichotomy) [72], [99], [96].

The notion of a Gaussian measure on a general locally compact topological group E [59] [29] [63] is much more complicated and for this reason has been less investigated than the corresponding notion on a linear space. Among the problems which arise in this comparatively new domain of probability theory, we mention the problem which is of main interest for us.

Problem. Let E be a locally compact separable group and μ and ν two Gaussian measures on E. Is it true that

$$\mu \sim \nu \text{ or } \mu \perp \nu?$$

Projectivity of Gaussian measures. Consider this problem in more detail and show that it can be included in some general scheme. That is, we are going to show that martingale theory [76] is the natural mathematical apparatus that could give us definite answers to this problem.

We fix a locally compact separable group E. In what follows we use the definition of a Gaussian measure given by Courrege [48] [63]. That is, a probability measure μ on E is said to be a Gaussian measure iff there exists a weakly continuous convolution semigroup $(\mu_t)_{t>0}$ of probability measures on E such that $\mu = \mu_1$ and

$$\lim_{t \to 0} \frac{1}{t} \mu_t(E \backslash U) = 0$$

holds for every neighborhood U of the identity $e \in E$.

It is clear that every measure μ_t is a Gaussian measure. According to [63, Th. 6.2.3], for every $t > 0$ the measure μ_t is supported on some closed subgroup of the component E_0 of the identity of E.

Thus when dealing with Gaussian measures it is sufficient to assume that the underlying locally compact group E is connected.

Now applying Glushkov's structural theorem, we see that E is a Lie projective group, i.e. there exists a descending sequence K_α of compact normal subgroups of E with $\cap K_\alpha = \{e\}$ such that $E_\alpha := E/K_\alpha$ is a Lie group for every α.

Now let $\mathcal{B}_\alpha$ be the σ-algebra of Borel subsets of E of the form $\pi_\alpha^{-1}(A)$, $A \in E_\alpha$, where π_α is the canonical mapping of E onto E_α. It is clear that $\mathcal{B}_\alpha \uparrow$ and $\sigma(\cup \mathcal{B}_\alpha) = \mathcal{B}$ where $\mathcal{B}$ is the Borel σ-algebra of E. For every α put $\mu_\alpha := \mu \mid \mathcal{B}_\alpha$ and note that μ_α can be considered as a Gaussian measure on the Lie group E_α.

Thus we see that every Gaussian measure μ on E is a projective limit of some projective sequence of Gaussian measures μ_α such that every measure μ_α is defined on the Lie group E_α.

Now let ν be another Gaussian measure on E with the associated projective sequence ν_α. It is clear that if $\mu \sim \nu$ then the following assumption holds:

Assumption (A): $\mu_\alpha \sim \nu_\alpha$ for every α.

Moreover, if $\mu_\alpha \perp \nu_\alpha$ for some α, then surely $\mu \perp \nu$.
Thus the problem becomes nontrivial if we suppose that assumption (A) holds.

Martingale background of the problem. The reasons from above show that the problem can be included in the following general scheme [67], [96].

Let $(\Omega, \mathcal{B}, \mathcal{B}_\alpha)$ be a measurable space with increasing family of σ-algebras $\mathcal{B}_\alpha$ such that $\mathcal{B}_\alpha \subset \mathcal{B}$ and $\sigma(\cup \mathcal{B}_\alpha) = \mathcal{B}$. Let us suppose that two probability measures P and $\widetilde{P}$ are given on $(\Omega, \mathcal{B})$ and let $P_n := P \mid \mathcal{B}_n$ and $\widetilde{P}_n := \widetilde{P} \mid \mathcal{B}_n$. Suppose that for every $n \geq 1$ the measure $\widetilde{P}_n$ is absolutely continuous with respect to the measure P_n. We write

$$z_n := d\widetilde{P}_n / dP_n,$$

the Radon–Nikodym derivative of $\widetilde{P}_n$ with respect to P_n.

It is clear that with respect to P the stochastic sequence $Z = (z_n, \mathcal{B}_n)_{n \geq 1}$ is a martingale. Consider the random value $z_\infty = \lim_{n \to \infty} z_n$ which exists P-a.s. The key to the problem on absolute continuity and singularity is Lebesgue's decomposition [67], [96]

$$\widetilde{P}(A) := \int_A z_\infty \, dP + \widetilde{P} \left(A \cap \{z_\infty = \infty\} \right)$$

According to this decomposition the question is reduced to the investigation of the probability of the *tail* event $\{z_\infty < \infty\}$, where one can use propositions like the Kolmogorov *zero-one* law.

Dichotomy problem for infinite products. A general test for absolute continuity or singularity can be given in terms of the sequence $q_n := z_n/z_{n-1}$ (see [67] and [90, 91, 92]). That is, note that

$$z_n := \prod_{k=1}^{n} q_k, \quad n \geq 1$$

but contrary to the case of "Kakutani dichotomy" [96], in general the random variables q_n are not independent, so we cannot use the Kolmogorov *zero-one* law. As it was shown in [90, 91, 92] the strong independence assumptions underlying Kakutani's theorem can be essentially weakened. That is, assuming that q_n is a sequence of uncorrelated functions, one can prove some kind of Kolmogorov *zero-one* law and the dichotomy theorem corresponding to this assumption.

An application of the technique of infinite products of uncorrelated functions to the dichotomy problem on a group seems to be useful and could give some definite results.

Gaussian measures on a linear space and on a group. Let E be a Lie projective group. It is known that there is a one-to-one correspondence between the set of Gaussian measures on E and the set of Gaussian measures on an appropriate linear space E^* [63] [59]. Does this means that, using this correspondence, we can prove that the dichotomy property on the group E is just a consequence of the dichotomy property on the linear space E^*? We consider here an example which serves to show that in general the answer to this question is NO!

Let $E := \mathbb{T}^\infty$ be the infinite dimensional torus, i.e. countable product of copies of the compact abelian group $\mathbb{T} = \mathbb{R}^1/2\pi\mathbb{Z}$. As we have already seen, Gaussian measures on $\mathbb{T}^\infty$ admit a very simple description. That is, let μ be a symmetric Gaussian measure and let $\widehat{\mu}(\theta)$ be its Fourier transform. According to [29] there exists a non-negative definite quadratic form $\psi(\theta)$ on the dual group $\mathbb{Z}^{(\infty)}$ such that

$$\widehat{\mu}(\theta) := \exp\{-\psi(\theta)\}, \quad \theta \in \mathbb{Z}^{(\infty)}.$$

In the natural basis of the discrete group $\mathbb{Z}^{(\infty)}$ this form can be written

$$\psi(\theta) = \sum_{i,j=1}^{\infty} a_{ij}\theta_i\theta_j, \quad \theta = (\theta_i) \in \mathbb{Z}^{(\infty)},$$

where the matrix $\mathscr{A} = (a_{ij})_{i,j=1}^{\infty}$ is *non-negative definite* in the sense that for every $n \geq 1$ the finite dimensional matrix $\mathscr{A}_n = (a_{i,j})_{i,j=1}^{\infty} \geq 0$.

Conversely, every *non-negative definite* matrix $\mathscr{A} = (a_{ij})_{i,j=1}^{\infty}$ generates in the above-mentioned sense some symmetric Gaussian measure $\mu^{\mathscr{A}}$ on $\mathbb{T}^{\infty}$.

Similar arguments show that there is a one-to-one correspondence $\mathscr{A} \Longleftrightarrow \widetilde{\mu}^{\mathscr{A}}$ between the set of *non-negative definite* matrices and the set of Gaussian measures on the linear space $\mathbb{R}^{\infty}$.

Thus the above-mentioned one-to-one correspondence $\mu^{\mathscr{A}} \Longleftrightarrow \widetilde{\mu}^{\mathscr{A}}$ between Gaussian measures on the group $\mathbb{T}^{\infty}$ and on the linear space $\mathbb{R}^{\infty}$ is established. This correspondence has the following form:

$$\mu^{\mathscr{A}} = \pi(\widetilde{\mu}^{\mathscr{A}})$$

where π is the canonical mapping of $\mathbb{R}^{\infty}$ onto $\mathbb{T}^{\infty} = \mathbb{R}^{\infty}/2\pi\mathbb{Z}^{\infty}$.

Example. Consider a sequence $(a_k)_{k\geq 1}$ of positive numbers such that the series $\sum_{k\geq 1} 1/a_k$ converges and let $\mathscr{A} = \mathrm{diag}\,(a_k)$ be a diagonal matrix. It can be shown that measures $\widetilde{\mu}^{\mathscr{A}}$ and $\widetilde{\mu}^{2\mathscr{A}}$ on $\mathbb{R}^{\infty}$ are singular (see [72] [96] [99]).

Now let $\mu^{\mathscr{A}}$ and $\mu^{2\mathscr{A}}$ be the corresponding Gaussian measures on the group $\mathbb{T}^{\infty}$. According to our assumption the measures $\mu^{\mathscr{A}}$ and $\mu^{2\mathscr{A}}$ are absolutely continuous with respect to Haar measure on $\mathbb{T}^{\infty}$ and have continuous and strictly positive densities $\mu^{\mathscr{A}}(x)$ and $\mu^{2\mathscr{A}}(x)$. Thus the measures $\mu^{\mathscr{A}}$ and $\mu^{2\mathscr{A}}$ are equivalent and the corresponding density $\mu^{\mathscr{A}}(x)/\mu^{2\mathscr{A}}(x)$ is a continuous and strictly positive function.

7.2 Harmonic functions on a group

The classical potential theory in $\mathbb{R}^n$ is translation-invariant; therefore it is natural to consider the harmonic space $(E, \mathscr{H})$ on a topological group E with a translation-invariant sheaf $\mathscr{H}$ of harmonic functions on E. The corresponding theory (the harmonic groups) was developed by Bliedtner [36] in 1968.

Translation-invariant elliptic operators or parabolic second-order differential operators on Lie groups were at that time the only source of examples of harmonic groups. In [37] Bliedtner conjectured that the underlying space of a harmonic group is necessarily a Lie group.

As we have already seen in Chapter 4, *the only abelian non-Lie groups that admit a potential theory in the sense of Bliedtner are groups of the form $\mathbb{R}^n \times \mathbb{T}^{\infty}$.*

Moreover, it was shown that every translation-invariant Brelot harmonic sheaf $\mathcal{H}$ on the group $\mathbb{R}^n \times \mathbb{T}^\infty$ can be described in terms of some elliptic second-order differential operator of infinitely many variables. That is, there exist a *non-negative definite* matrix $\mathcal{A} = (a_{ij})_{i,j=1}^\infty$ and a vector $b = (b_i)_{i=1}^\infty$ defined uniquely up to a multiplicative constant and such that every harmonic function u is a weak solution of the equation

$$\sum_{i,j=1}^\infty a_{ij}\partial_i\partial_j u(x) + \sum_{i=1}^\infty b_i\partial_i u(x) = 0, \quad x = (x_i) \in \mathbb{R}^n \times \mathbb{T}^\infty.$$

The nonabelian case is still open, i.e. there is no description of Brelot harmonic sheaves on a general locally compact group E.

There is no doubt that there is a close connection between the topological structure of a group E and properties of Gaussian measures and harmonic functions on it.

7.3 The problem of hypoellipticity

Let $\mathcal{H}$ be a translation-invariant Brelot harmonic sheaf on the group $E = \mathbb{T}^\infty$. As mentioned above, there exists an infinite-dimensional elliptic operator $\mathcal{L} = \sum a_{ij}\partial_i\partial_j + \sum b_i\partial_i$ such that every harmonic function u is a weak solution of the equation $\mathcal{L}u = 0$. *Does it follow that u is a strong solution?*

In the case $\mathcal{L} = \sum a_k\partial_k^2$ the exhaustive answer to this problem can be given in terms of the counting function $N(\lambda) = \sum_{a_k \leq \lambda} 1$ of the coefficients of the elliptic operator $\mathcal{L}$ (see Theorem 2.3.3); some particular results concerning the nondiagonal case can be found in [24].

The following two properties are equivalent:

1) *Hypoelliptic property*: every weak solution of the equation $\mathcal{L}u = 0$ is a strong solution,
2) $N(\lambda) = o(\lambda)$ as $\lambda \to \infty$.

We note two crucial moments in the proof of this result which illustrate the close connection between properties of harmonic functions and Gaussian measures on a group.

1. Let $(P_t)_{t>0}$ be the symmetric Markov semigroup corresponding to the operator $\mathcal{L}$. Standard Fourier transform arguments show that for every $t > 0$ we have $P_t f = \mu^{t\mathcal{A}} * f$ where $\mu^{t\mathcal{A}}$ is the Gaussian measure corresponding to the matrix $t\mathcal{A} = \operatorname{diag}(ta_k)$. The operator $\mathcal{L}$ admits the hypoelliptic property iff the following properties hold:

(i)   for every $t > 0$ the Gaussian measure $\mu^{t\mathcal{A}}$ is absolutely continuous with respect to Haar measure and has continuous density $\mu^{t\mathcal{A}}(x)$;

(ii) for every $\varepsilon > 0$ we have

$$\lim_{t \to 0} e^{-\varepsilon/t} \mu^{t\mathcal{A}}(0) = 0.$$

2. There exists a strictly positive function $k : \mathbb{R}^1_+ \longmapsto \mathbb{R}^1_+$ such that

$$\log \mu^{t\mathcal{A}}(0) = k \star N\left(\frac{1}{t}\right),$$

where $\star$ is the Mellin convolution on the multiplicative group $(\mathbb{R}^1_+, \cdot)$ (see [27]).

In the noncommutative case the following problem seems to be of interest. Fix some sequence E_k of compact Lie groups and consider a Lie projective compact group $E = \prod_{k=1}^{\infty} E_k$. On every E_k let the invariant metric d_k and the corresponding Laplacian $\mathcal{L}_k$ be given. Consider the Laplace operator $\mathcal{L} = \sum_{k=1}^{\infty} \mathcal{L}_k$ on the group E.

Problem. Find conditions on the sequence (E_k, d_k) such that the operator $\mathcal{L}$ is hypoelliptic.

7.4 "Can one hear the shape of a drum?"

This question is the title of a famous article by Mark Kac which appeared in 1966 (Amer. Math. Monthly, **73**, pp. 1–23). In this landmark paper, Kac showed that geometric properties of regions in $\mathbb{R}^2$ can be obtained by studying the asymptotic properties of the spectrum of the Laplacian.

Admittedly imprecise as a mathematical problem, this question belongs to the increasingly large field of "inverse problems." These are more familiar in the case of inverse spectral problems in which one seeks to recover geometric information beginning with the eigenvalues of a differential operator [50], [87], [89].

We start with the group $E = \mathbb{T}^{\infty}$ and with the differential operator $\mathcal{L} = \sum a_k \partial_k^2$ defined by the sequence a_k.

Let Ω be an open subset of E and $\mathcal{L}_\Omega$ be the restriction of the operator $\mathcal{L}$ to Ω subject to the Dirichlet boundary conditions.

Suppose that $\mathcal{L}_\Omega$ has a discrete spectrum $\{\lambda_k(\Omega)\}$ and suppose that we know the behavior of its eigenvalues

$$\lambda_k(\Omega) \sim \Lambda(k, \Omega), \quad k \to \infty.$$

Could we derive from this assumption some information about the behavior of the sequence $\{a_k\}$ of the coefficients of the operator $\mathcal{L}$?

Under some natural assumptions the answer to this question is YES!

We give here only a sketch of our reasons, which once again shows the importance of the investigation of Gaussian measures on a group.

As above, let $(\mu^{t\mathscr{A}})_{t>0}$ be the Gaussian semigroup on $\mathbb{T}^\infty$ associated with the operator $\mathscr{L}$. Suppose that for every $t > 0$ the measure $\mu^{t\mathscr{A}}$ has a continuous density $\mu^{t\mathscr{A}}(x)$ with respect to the Haar measure. It follows that for every domain $\Omega \subseteq \mathbb{T}^\infty$ the operator $\mathscr{L}_\Omega$ has a discrete spectrum $\{\lambda_k(\Omega)\}$. Let $|\Omega|$ be the Haar measure of the domain Ω. Consider the following function:

$$M_\Omega(\lambda) := \frac{1}{|\Omega|} \sum_{\lambda_k(\Omega) \leq \lambda} 1$$

It can be shown that the following asymptotic equality holds:

$$\log \int_0^\infty e^{-\lambda t} dM_\Omega(\lambda) \sim \log \mu^{t\mathscr{A}}(0) \text{ as } t \to 0$$

The next essential step in our reasons is the identity

$$\log \mu^{t\mathscr{A}}(0) = k \star N\left(\frac{1}{t}\right),$$

which was already used in Sec. 4.3.

Thus finally we find

$$\log \int_0^\infty e^{-\lambda t} dM_\Omega(\lambda) \sim k \star N\left(\frac{1}{t}\right) \text{ as } t \to 0$$

The following particular result is just an application of the Kohlbecker Tauberian theorem to the asymptotic equality from above.

Let $\alpha > 0$. The following properties are equivalent:

(i) $a_k \sim k^\alpha$ as $k \to \infty$;

(ii) $\lambda_k(\Omega) \sim c_\alpha \left(\log \frac{k}{|\Omega|}\right)^{1+\alpha}$ as $k \to \infty$.

The following problem seems to be of interest. Let $\{E_k\}$ be some sequence of compact Lie groups. Suppose that on each E_k an invariant metric d_k is given and let $\mathscr{L}_k$ be the Laplacian on E_k corresponding to d_k. Consider a compact group $E = \prod_{k=1}^\infty E_k$ and infinite dimensional Laplacian $\mathscr{L} = \sum_{k=1}^\infty \mathscr{L}_k$ on E.

Problem. What kind of information about the sequence $\{d_k\}$ can we derive from knowledge of the spectrum of the operator $\mathscr{L}$ on E?

7.5 Geometry on a group

Let $\{E_k\}$ be a sequence of compact Lie groups and $E = \prod_{k=1}^\infty E_k$. Suppose that on each E_k an invariant metric d_k is defined and let $\mathscr{L}_k$, ∇_k, and dx_k be the Laplacian, the gradient, and the normed Riemanian volume on E_k corresponding to d_k.

Put $\mathcal{L} = \sum_{k=1}^{\infty} \mathcal{L}_k$; $dx = \otimes_{k=1}^{\infty} dx_k$; and let $(\mu_t)_{t>0}$ be the Gaussian convolution semigroup corresponding to $\mathcal{L}$. Define the following function on $E \times E$:

$$d(x, y) := \sup\{\psi(x) - \psi(y),\ \psi \in \mathcal{D}^1,\ |\nabla\psi|^2 \le 1\}$$

where $\mathcal{D}^1$ is the set of smooth cylindrical functions on E and $|\nabla\psi|^2 = \sum_{k=1}^{\infty} |\nabla_k\psi|^2$.

In general $d(x, y)$ is a pseudometric (i.e. $d(x, y) = \infty$ for some $x \neq y$) [35] [25] [105]. The following theorem seems to be true.

Theorem. The following statements are equivalent:
(i) $d(x, y)$ is a complete metric on E which is compatible with the original topology on E;
(ii) for every $t > 0$ the measure μ_t has a continuous density $\mu_t(x)$ with respect to the Haar measure dx on E and

$$\lim_{t\to 0} -4t \log \mu_t(x \cdot y^{-1}) = d^2(x, y)$$

holds for all $x, y \in E$.

At least in the case $E = \mathbb{T}^\infty$ and $\mathcal{L} = \sum_{k=1}^{\infty} a_k \partial_k^2$, the equivalence (i) $\Leftrightarrow$ (ii) has been proved in a forthcoming paper [27].

Bibliography

[1] Bauer, H. *Harmonische Räume und ihre Potentialtheorie.* Lect. Notes in Math. **22**. Berlin-Heidelberg-New York: Springer Verlag (1966).

[2] Bauer, H. *Harmonic spaces and associated Markov processes.* In: Potential Theory, C.I.M.E. (Stresa, 1969), Rome (1970).

[3] Bauer, H. *Elliptische harmonische Strukturen.* In: Elliptische Differentialgleichungen (Rostock, 1977). Math. Ges. der DDR (1978).

[4] Bauer, H. *Elliptic differential operators and diffusion processes.* Bull. Austral. Math. Soc., **30** (1984), 219 –237.

[5] Bauer, H. *Harmonic Spaces - A Survey.* In: Comp. del. Semin. di Mat. dell'Université de Bari, 1 –26. Bari (1984).

[6] Bendikov, A. D. *Space-homogeneous continuous Markov processes on abelian groups and harmonic structures.* Uspekhi Mat. Nauk, **29**, no. 5 (179) (1974), 215 – 216. (=Russian Math. Surveys 29:5 (1974), 215 –216)

[7] Bendikov, A. D. *Functions that are harmonic for a class of diffusion processes on a group.* Teor. Veroyatn. i Primen., **20**, no. 4 (1975), 773 –784. Theory Prob. Appl. **20** (1975), 759 –769.

[8] Bendikov, A. D. *A continuity criterion for a class of Markov processes.* Teor. Veroyatn. i Primen., **21** (1976), 169 –171. (=Theory Probab. Appl. 21 (1976), 169 –171)

[9] Bendikov, A. D. *Harmonic functions for a category of Markov processes and projective limits in it.* Uspekhi Mat. Nauk, **31**, no. 2 (1976), 209 –210. (=Russian Math. Surveys 31:2 (1976), 209 –210)

[10] Bendikov, A. D. *Harmonic structures generated by a Wiener heat process on a group.* Uspekhi Mat. Nauk, **34**, no. 1 (205) (1979), 217 –218. (=Russian Math. Surveys 34:1 (1979), 217 –218)

[11] Bendikov, A. D. *Harmonic spaces generated by Markov processes.* VINITI (1979), No. 5:135.

[12] Bendikov, A. D., and Pavlov, I. V. *Spaces H^p and BMO on the infinite dimensional torus.* III International Vilnius Conference on Prob. Theory and Math. Statistics, Vilnius, **1** (1981), 26 –27.

[13] Bendikov, A. D., and Pavlov, I. V. *A duality theorem for function spaces H^1 and BMO on the infinite-dimensional torus.* Random Analysis and Assymptotic Problems in Probability Theory and Mathematical Statistics. Tbilisi, Metsniereba (1984), 6 – 16.

[14] Bendikov, A. D. *Diffusion processes and partial differential equations on the group.* I International Congress of the Bernoulli Society of Mathematical Statistics and Probability Theory, **2**. Moscow, Nauka (1986), 602 –603.

[15] Bendikov, A. D., and Pavlov, I. V. *Boundedness on $L_p(\mathbb{T}^\infty)$ of a class of vector multiplier operators.* Sib. Mat. Zhurn., **27**, no. 1 (1986), 3 –10. (=Siberian Math. J. 27:1 (1986), 1 –7)

[16] Bendikov, A. D., and Pavlov, I. V. *Diffusion processes and elliptic equations of infinitely many variables.* In: Probability Theory and Math. Stat. Proceedings of the Fourth Vilnius Conference, **1** (1987), 145 –170. VNU Science Press BV.

[17] Bendikov, A. D., and Pavlov, I. V. *Spaces $L_{\vec{p}}$ with mixed norm on infinite Cartesian product of probability spaces.* Analysis Math., no. 13 (1987), 231 –250.

[18] Bendikov, A. D. *Markov processes and partial differential equations on a group: the space-homogeneous case.* Uspekhi Mat. Nauk, **42**, no. 5 (257) (1987), 41 –78. Russian Math Surveys, **42**:5 (1987), 49 –94.

[19] Bendikov, A. D., and Pavlov, I. V. *Spaces of harmonic functions with martingale mixed norm.* Teoriya Veroyatnostei i Primen., **33** (1988), 769 –772. (= Theory Probab. Appl. 33 (1988), 769 –772)

[20] Bendikov, A. D., and Pavlov, I. V. A *Wiener process with reflection and harmonic functions with finite energy integral.* Teoriya Veroyatn. i Primen., **33** (1988), 586 – 589. (= Theory Probab. Appl. 33 (1988), 586 –589)

[21] Bendikov, A. D. *Markov processes and harmonic structures.* Statistics and Control of Stochastic Processes, A. N. Shiryaev et al., ed., Moscow, Nauka (1989), 25 –32.

[22] Bendikov, A. D. *On Markov processes associated with projective sequences of harmonic spaces.* Proc. of V International Conference on Probability Theory and Mathematical Statistics, VSP/MOKSLAS, Utrecht/Vilnius, **1** (1990), 118 –128.

[23] Bendikov, A. D. *Some properties of Gaussian measures and potentials in martingale space with mixed norm.* New Trends in Probability and Statistics, V. V. Sazonov and T. Shervashidze (eds.), VSP/MOKSLAS, Utrecht/Vilnius, **1** (1991), 1 –12.

[24] Bendikov, A. D. *Uniformly elliptic operators on the infinite dimensional torus.* C. R. Acad. Sci., Serie I, **316** (1993), 733 –738.

[25] Bendikov, A. D. *Remarks on analysis on local Dirichlet spaces.* To appear in: Proc. ICDFSP, Beijing, China (1993). Berlin-New York: Walter de Gruyter (1995).

[26] Bendikov, A. D. *On the problem of the hypoellipticity on the infinite dimensional torus.* Proc. NATOARW "Classical and Axiomatic Potential Theory." (1994)

[27] Bendikov, A. D. *Symmetric stable semigroups.* Exposition. Math., 1995.

[28] Benedek, A., and Panzone, R. *The spaces L_p with mixed norm.* Duke Math. J., no. 28 (1961), 301 –324.

[29] Berg, C., and Forst, G. *Potential Theory on Locally Compact Abelian Groups*, Lect. Notes in Math. **87** (1975). Berlin-Heidelberg-New York: Springer Verlag.

[30] Berg, C. *Potential theory on the infinite dimensional torus*. Invent. Math., **32** (1976), 49 –100.

[31] Berg, C. *On the support of the measures in a symmetric convolution semigroup*. Math. Z., **148** (1976), 141 –146.

[32] Berg, C. *On Brownian and Poissonian convolution semigroups on the infinite dimensional torus*. Invent. Math., **38** (1977), 227 –235.

[33] Berg, C. *Hunt convolution kernels which are continuous singular with respect to Haar measure*. In: Probability measures on groups. Proc. Oberwolfach, Lect. Notes in Math. no. 706 (1978). Berlin-Heidelberg-New York: Springer Verlag.

[34] Besov, O. V; Il'in, V. P.; and Nikol'skii, S. M. *Integral Representations of Functions and Imbedding Theorems*. Moscow: Nauka (1975).

[35] Biroli, M., and Mosco, U. *A Saint-Venant Principle for Dirichlet forms on discontinuous media*. Elenco preprint, Rome (1992).

[36] Bliedtner, J. *Harmonische Gruppen und Huntsche Faltungskernel*, Semin. über Potentialtheorie, Lect. Notes in Math. **69** (1968). Berlin-Heidelberg-New York: Springer Verlag.

[37] Bliedtner, J. *On the analytic structure of harmonic groups*. Manuscripta Math., **1** (1969), 289 –292.

[38] Bliedtner, J., and Hansen, W. *Markov processes and harmonic spaces*. Z. Wahrscheinlichkeistheorie, **42** (1978), 309 –325.

[39] Bliedtner, J., and Hansen, W. *Potential Theory: an Analytic and Probabilistic Approach to Balayage*. Universitext, Springer (1986).

[40] Blumenthal, R. M., and Getoor, R. K. *Markov Processes and Potential Theory*. New York, London: Academic Press (1968).

[41] Boboc, N.; Constantinescu, C.; and Cornea, A. *Semigroups of transitions of harmonic spaces*. Rev. Roumaine Math. Pures Appl., no. 12 (1967), 763 –805.

[42] Bony, J.-M. *Opérateurs elliptiques dégénérés associés aux axiomatiques de la théorie du potential*. In: Potential Theory. C.I.M.E. (Stresa, 1969), 69 –119. Rome (1970).

[43] Brelot, M. *Axiomatique des functions harmoniques*. Univ. de Montreal (1966).

[44] Brelot, M. *Éléments de la théorie classique du potentiel*. 4th ed. Centre de Documentation Universitaire Paris (1969).

[45] Brelot, M. *On Topologies and Boundaries in Potential Theory*. Lecture Notes in Math. **175**, Berlin, Springer-Verlag, 1971.

[46] Constantinescu, C.; and Cornea, A. *Potential Theory on Harmonic Spaces*. Lect. Notes in Math. **158** (1972). Berlin-Heidelberg-New York: Springer Verlag.

[47] Courrege, P., and Priouret, P. *Axiomatique du probleme de Dirichlet et processus de Markov*. In: Semin. de Theorie du Potentiel (Brelot, Choquet, Deny), Inst. Henri Poincare, 8 annee, 1963/64, No. 8, p. 48.

[48] Courrege, P. *Generateur infinitesimal d'un semi-groupe de convolution sur $\mathbb{R}^n$ et formule de Levy-Khinchine.* Bull. Sci. Math, 2^e Ser. 88 (1964), 3 –30.

[49] Daletskii, Yu. L., and Fomin, S. V. *Measures and Differential Equations in Infinite Dimensional Spaces.* Moscow: Nauka (1983).

[50] Dodziuk, J. *Eigenvalues of the Laplacian and the heat equation.* Amer. Math. Monthly, **88** (1981), 686 –695.

[51] Doob, J. L. *Semimartingales and sub-harmonic functions.* Trans. Amer. Math. Soc., **77** (1954), 86 –121.

[52] Doob, J. L. *A probability approach to the heat equation.* Trans. Amer. Math. Soc., **78** (1955), 216 –280.

[53] Doob, J. L. *Classical Potential Theory and its Probabilistic Counterpart.* Grundl. d. math. Wiss. **262** (1983). Berlin-Heidelberg-New York: Springer Verlag.

[54] Dynkin, E. B. *Markov Processes.* I, II. Grundl. d. math. Wiss. **121, 122**, Springer-Verlag (1965).

[55] Forst, G. *Symmetric harmonic groups and translation invariant Dirichlet spaces.* Invent. Math., **18**, No. 1 –2 (1972), 143 –182.

[56] Fukushima, M. *Dirichlet Forms and Markov Processes.* North-Holland Math. Library **23** (1980).

[57] Gikhman, I. I., and Skorokhod, A. V. *Theory of Random Processes.* Vol. 1. Moscow: Nauka (1971).

[58] Goodman, V. *Harmonic functions on Hilbert spaces.* J. Funct. Anal., no. 10 (1972), 451 –470.

[59] Grenander, U. *Probabilities on Algebraic Structures.* Stockholm-Göteborg-Uppsala: Almquist & Wiksell (1963).

[60] Guggenmoos-Holzmann, J. *Harmonische Funktionen auf Abstrakten Wienerräumen.* Den Naturwissenschaftlichen Fachbereichen der Friedrich-Alexander-Universität Erlangen-Nürnberg. Doctoral dissertation (1977).

[61] Hervé, R.-M. *Recherches axiomatiques sur la theorie du fonctions surharmoniques et du potentiel.* Ann. Inst. Fourier, **12** (1962), 415 –571.

[62] Hervé, R.-M. *Quelques propriétés des fonctions surharmoniques associés a'une équation uniformément elliptique de la forme $Lu = -\sum_i \frac{\partial}{\partial x_i}\left(\sum_j a_{ij}\frac{\partial u}{\partial x_j}\right) = 0$.* Ann. Inst. Fourier, no. 15/2 (1965), 215 –223.

[63] Heyer, H. *Probability Measures on Locally Compact Groups.* Ergeb. der Math. und ihren Grenzgeb., **94** (1977). Berlin-Heidelberg-New York: Springer Verlag.

[64] Holley, R., and Stroock, D. *Diffusion on an infinite dimensional torus.* J. Funct. Anal., **42** (1981), 29 –43.

[65] Hunt, J. A. *Markov Processes and Potentials.* I, II, III. Ill. J. Math. **1, 2, 3** (1957, 1957, 1959), 44 –93, 316 –369, 151 –213.

[66] Ito, K., and McKean, H. *Diffusion Processes and Their Simple Paths*. Academic Press, Inc.; Berlin-Heidelberg-New York, Springer-Verlag (1965).

[67] Kabanov, Yu.; Liptser, R. Sh.; and Shiryaev, A. N. *On the question of the absolute continuity and singularity of probability measures*. Math USSR-Sb., **33** (1977), 203 – 221.

[68] Kac, M. *Can one hear the shape of a drum?* Amer. Math. Monthly, **73** (1966), 1 –23.

[69] Kakutani, S. *On equivalence of infinite product measures*. Ann. of Math., **49** (1948), 214 –224.

[70] Kantorovich, L. V., and Akilov, G. P. *Functional Analysis*. Moscow: Nauka (1977).

[71] Krein, S. G.; Petunin, Yu. I.; and Semenov, E. M. *Interpolation of Linear Operators*. Moscow: Nauka (1978).

[72] Kuo, H. *Gaussian measures on Banach space*. Lecture Notes in Math. **463** (1975). Berlin-Heidelberg-New York: Springer Verlag.

[73] Landis, E. M. *Second-order Elliptic and Parabolic Equations*. Moscow: Nauka (1971).

[74] Landkof, N. S. *Foundations of Modern Potential Theory*. Grundl. d. math. Wiss. **180**, Springer-Verlag (1972).

[75] Lancaster, P. *Matrix Theory*. Moscow: Nauka (1978) (1969).

[76] Liptser, R. Sh., and Shiryaev, A. N. *Theory of Martingales*. Kluwer Acad. Publ. Vol. 49 (1989).

[77] McKean, H. *Stochastic Integrals*. New York-London: Academic Press (1969).

[78] Meyer, P.-A. *Brelot's axiomatic theory of the Dirichlet problem and Hunt's theory*. Ann. Inst. Fourier, **13**, no. 2 (1963), 357 –372.

[79] Meyer, P.-A. *Demonstration probabiliste de certaines inequalities de Littlewood-Paley*. In: Lecture Notes in Math. **511** (1976), 125 –183. Berlin-Heidelberg-New York: Springer Verlag.

[80] Meyer, P.-A. *Retoor sur la theorie de Littlewood-Paley*. Lecture Notes in Math. **850** (1981), 151 –166. Berlin-Heidelberg-New York: Springer Verlag

[81] Meyer, P.-A. *Probability and Potentials*. Blaisdell Publ. Comp. (1966).

[82] Mizohata, S. *Theory of Partial Differential Equations*. London, Cambridge Univ. Press (1973).

[83] Neveu, J. *Mathematical Foundations of Probability Theory*. Moscow: Mir (1969) (1964).

[84] Neveu, J. *Discrete-parameter Martingales*. North-Holland (1975).

[85] Parthasarathy, K. R. *Probability Measures on Metric Spaces*. New York-London: Academic Press (1967).

[86] Pavlov, I. V. *Probability-analytic approach to the investigation of Hardy and BMO spaces on the infinite-dimensional torus*. Dissertation, Vilnius State University (1982).

[87] Pinsky, M. *Can you feel the shape of a manifold with Brownian motion?* Exposition. Math., **2** (1984), 289 –298.

[88] Port, S. C., and Stone, C. I. *Infinite divisible processes and their potential theory*. Ann. Inst. Fourier, **21**, no. 2 (1971), 157 –275 (Part I); no. 4, 179 –265 (Part II).

[89] Protter, M. H. *Can one hear the shape of a drum? Revisited*. SIAM Review, **29**, no. 2 (June 1987), 185 –197.

[90] Ritter, G. *Unendliche Produkte unkorrelierter Funktionen auf kompakten, abelschen Gruppen*. Math. Scand., **42** (1978), 251 –270.

[91] Ritter, G. *On dichotomy of Riesz products*. Math. Proc. Camb. Phil. Soc., **85** (1979), 79 –89.

[92] Ritter, G. *On Kakutani's dichotomy theorem for infinite products of not necessarily independent functions*. Math. Ann., **239** (1979), 35 –53.

[93] Schirmeier, U. *Produkte harmonischer Räume*. Sitzungsber. Bayer. Akad. Wiss. Math.-Natur. Kl., 1978 (1979), 5 –22.

[94] Schirmeier, U. *Konvergenzeigenschaften in harmonischen Räumen*, Inventiones math. **55** (1979), 71 –95.

[95] Schmidt, K. *Unique ergodicity for quasi-invariant measures*. Preprint, February 1978, from Math. Inst. University of Warwick, Coventry.

[96] Shiryaev, A. N. *Probability*. Grad. Texts in Math. **95** (1984). Berlin-Heidelberg-New York: Springer Verlag.

[97] Shur, M. G. *Harmonic functions for a Markov process*. Mat. Zametki, **13**, no. 4 (1973), 587 –596.

[98] Silverstein, M. L. *Application of the sector condition to the classification of submarkovian semigroups*. Trans. Amer. Math. Soc., **244** (1978), 103 –146.

[99] Skorokhod, A. V. *Integration in Hilbert Space*. Berlin-Heidelberg-New York: Springer Verlag (1974).

[100] Sondermann, D. *Mase auf lokalbeschränkten Räumen*. Ann. Inst. Fourier, **19**, no. 2 (1970), 32 –113.

[101] Stein, E. M. *Singular Integrals and Differential Properties of Functions*. Princeton, New Jersey, Princeton University Press (1971).

[102] Stein, E. M., and Weis, G. *Introduction to Harmonic Analysis on Euclidean Spaces*. Princeton, New Jersey, Princeton University Press (1971).

[103] Stein, E. M. *Some results in harmonic analysis in $\mathbb{R}^n$, for $n \to \infty$* . Bull. Amer. Math. Soc., **9**, no. 1 (1983), 71 –73.

[104] Stoica, L. *Local Operators and Markov Processes.* Lect. Notes in Math. **816** (1980). Berlin-Heidelberg-New York: Springer Verlag.

[105] Sturm, K.-T. *On the geometry defined by Dirichlet forms.* Preprint, Erlangen (1993).

[106] Tautz, G. *Zur Theorie der ersten Randwertaufgabe.* Math. Nachr., no. 2 (1949), 279 –303.

[107] Taylor, J. C. *Duality and the Martin compactification.* Ann. Inst. Fourier, no. 22/3 (1972), 95 –130.

[108] Taylor, J. C. *The harmonic spaces associated with "reasonable" standard processes.* Math. Ann., **233**, no. 1 (1978), 89 –96.

[109] Yosida, K. *Functional Analysis.* Berlin: Springer (1968).

[110] Zabreiko, P. P. *Ideal Spaces of Functions.* Vestnik Yaroslavsk. Univ., no. 8 (1974).

Index